George Goudie Chisholm

Animal Life on the Globe

George Goudie Chisholm

Animal Life on the Globe

ISBN/EAN: 9783744670784

Printed in Europe, USA, Canada, Australia, Japan

Cover: Foto ©berggeist007 / pixelio.de

More available books at **www.hansebooks.com**

ANIMAL LIFE ON THE GLOBE

BY

G. G. CHISHOLM, M.A., B.Sc., F.R.G.S.

A Geographical Reading Book, describing some Typical Animals and
Illustrating the Relation of their Habits to the Character
of the Countries in which the animals are
respectively found, as well as the
broader distinctions of
Animal Life

———

BOSTON
BOSTON SCHOOL SUPPLY COMPANY
15 BROMFIELD STREET
1895

TYPOGRAPHY BY C. J. PETERS & SON,
BOSTON.

CONTENTS.

5

ANIMAL LIFE.

LESSON I.

far'-thest	ea'-si-ly	stop'-ped	spread'-ing
al'-ways	win'-ter	mid'-dle	to-geth'-er
gar'-ment	warmth	bur'-i-ed	bar'-ley

THE PART OF THE WORLD ROUND THE NORTH POLE.

1. You know by this time where to find the North Pole, when you look at a globe. The North Pole on a globe stands for the point farthest north on this round world. If you could reach that point, you could not go a step away from it, without going back again towards the south.

2. But no one ever has reached that spot on the earth; and if you were to go as far north as you could, you would easily see why the North Pole has never been reached by man.

3. You would find that, long before you got near the Pole, you were stopped by vast fields of ice and snow spreading as far as you could see.

4. The cold would become so great that you could not bear it, and those who have tried to go over the ice to the North Pole have always found that they could not go very far.

5. Even in the lands that are not so near the North Pole, many things are quite unlike what they are in our own land. Very strange it would seem to any boy or girl, who could spend a year in one of them.

6. For in these lands the winter lasts eight or nine months, and in the middle of the winter there is no day at all. For weeks together the sun is never seen, and the light of the stars shows only the white garment of deep snow, in which all the earth is wrapt.

7. Before and after this long night, the days are very short, but they grow longer

and longer when the summer is near. At last the people in those lands enjoy longer summer days than we ever know at home, and during the middle of the summer there is no night, just as in the middle of the winter there is no day.

8. Under the warmth of the sun during these long days, the snow quickly melts away. In a very short time the whole face of the land is changed. Green plants and bright flowers spring up, where the ground was so long buried under snow.

9. So quickly do plants grow, that in some parts of these countries, crops of barley can be sown and reaped within about two months.

10. You may be sure, that in these northern lands, the animals must be very unlike any we have at home. Think for a moment what would become of our sheep and cattle there. They could not stand the cold, and if they could, they would find no food to keep them alive through the long winter.

11. Yet there are many animals that dwell in these countries, and in the next few lessons you will read about some of these, and learn how they are fitted to thrive in a part of the world where our animals would die.

——— ◂•▸ ———

LESSON II.

height	an′–i–mals	pic′–ture	shoul′–ders
ea′–si–er	her–biv′–o–rous	suck′–led	moth′–er
hoofs	cu′–ri–ous	sur′–face	more′–over
broad	rein′–deer	mam′–mal	spread

THE REINDEER. 1.

1. THE reindeer is the first of the animals spoken of in the last lesson, about which you will have to read. The picture on page 14 shows you what it is like. Its height at the shoulders is nearly four feet.

2. Now that you know what sort of a country it lives in. you will be at a loss to think what kind of food it eats. You will feel sure that it cannot feed like our own

deer on grass. No; it does not feed on grass, yet it does feed wholly on herbs or plants, or things got from plants. It is for that reason called herbivorous, which simply means herb-eating.

3. The young reindeer does not feed on herbs, however. It is suckled by its mother with milk, and because it does so, the reindeer is called a mammal.

4. There are a great many animals that feed their young in the same way. All of these are mammals, and the reindeer, like our own deer, and like the sheep or cow, is called an herbivorous mammal, because when full-grown it feeds on herbs.

Moose.

5. What sort of herb does the reindeer feed on? It is a very curious one, and very

unlike the grass on which our cattle feed. It is one that is not killed by the winter cold, or spoiled by the snow, and because the reindeer feeds on it, it is called the reindeer moss.

6. This plant has no stem or leaves. It looks more like a sponge than anything else, except that its color is a greenish white.

7. In some way which we cannot explain, the reindeer knows where to find this plant, even when buried deep under the snow, and on coming to such a spot, it begins to scrape away the snow with its forefeet, until it reaches the moss.

8. Now you know how the reindeer is able to live in countries so strange as those of which you read in the first lesson. But there are other things also, that make this animal well fitted for the life it leads.

9. Its feet are so made, that it is easier for it to run on the snow than it would be for an animal like a cow. Hoofs cover the toes, as they do in the cow, but the hoofs are

very broad, and the feet are split between
the hoofs pretty high up, so that when they
touch the ground, they spread out over a
large surface. In that way they help to keep
the animal from sinking deep in the snow.

10. Moreover, there are long hairs under
the feet, and these by getting caught in the
snow have the same use. In summer as well
as in winter, this kind of foot is of use to the
reindeer, for when the snow is all melted away,
its place is taken by soft marshy ground.

LESSON III.

peo′-ple	use′-ful	Nor′-way	Swe′-den
islands	blank′-ets	sup-ply′	al′-most
freez′-ing	glue	sin′-ews	fro′-zen
thread	leath′-er	weigh′-ing	cloth′-ing

THE REINDEER. 2.

1. To the people who dwell in those icy
lands in the far north, where reindeer herds
are found, this animal is so useful that many
of them could not live without it.

2. The people who make most use of it are the Lapps, whose home is in the north of Europe, between the White Sea and At-

lantic Ocean, north of the Arctic Circle, and is called Lapland.

3. By the Lapps the reindeer has been tamed, and it is kept by them in large flocks,

like sheep or cows among us. In a country where little else is to be had, the Lapps get from the reindeer almost all that they need. They get food and clothing and things to sell to the people of other countries, so that they may have money to buy what their own country does not supply them with.

4. Almost every part of the reindeer is made use of. The milk, blood, and flesh are all used as food. The milk is very thick and rich, much thicker and richer than that of the cow.

5. The flesh is good to eat, and the Lapps can keep it for any length of time in winter by freezing it.

6. Frozen reindeer meat is one of the things that they sell to the people, who live where there are no reindeer. They also sell the hoofs and branching horns to make glue. Out of the sinews of the reindeer thread is made. The skin is made into a very good leather.

7. But the reindeer has other uses also. In summer it carries loads strapped across

its back, and in winter it is much more useful, because it can then draw sledges across the snow. The sledges are shaped something like the front half of a boat, very sharp in front, so that they can be drawn along very easily.

8. A sledge like that with a man in it, or with a load weighing nearly two hundred pounds, can be drawn by a single reindeer over the snow, at the rate of about nine miles an hour. As the reindeer does not easily get tired, but can keep up this speed for many hours, a Lapp may travel more than a hundred miles in one day.

9. Even in winter the Lapps have to travel a great deal, for they must always be near the food of the reindeer, and when the reindeer moss is nearly eaten up at one place, they must move to another. Hence they do not live in proper houses, but, cold as their country is, they live in tents, which they can carry about with them.

10. By clothing themselves in the skins of the reindeer, which are covered with long

hair or fur, and by using these skins for bedding and blankets, they manage to keep themselves warm enough.

———————

LESSON IV.

suck′–les	her–biv′–o–rous	ea′–si–er	ca′–nine
tru′–ly	car–niv′–o–rous	po′–lar	ter′–ri–ble
pieces	length	weight	no′–tice
in–stead′	seiz′–ing	up′–per	mam′–mals

THE POLAR BEAR. — 1.

1. THE polar bear, like the reindeer, suckles its young with milk, and is therefore a mammal. But it is not an herbivorous mammal, as the reindeer is, for it does not feed on plants, or on anything got from plants. It feeds on flesh only, and so it is a carnivorous mammal.

2. Carnivorous means just the same as the easier word flesh-eating. Dogs and cats, as you know, eat flesh, and if they were wild and lived as they liked, they would eat flesh only, and then they too would be truly carnivorous animals.

3. If you look into a dog's mouth, you may see four sharp-pointed teeth, longer than the rest, one at each side of the front,

in both the upper and lower jaw. These teeth are called canines or dogs' teeth. Now, the polar bear has very long and terrible canines. Indeed, all carnivorous animals have canines. Can you not tell why?

4. Because they need teeth to tear in pieces the flesh of the animals on which they feed. The teeth of the reindeer are very unlike the teeth of the polar bear, for they are never used to tear flesh.

5. The polar bear is larger than the reindeer. It is, in fact, the largest of all carnivorous mammals that live on land. Sometimes a full-grown male is eight and a half feet in length from nose to tail, and its weight may be as much as eight hundred pounds. A polar bear of that size in one scale would weigh down five men in the other scale.

6. Look at the pictures of the polar bear and the reindeer, and notice how different the two animals are in other ways. The polar bear has no horns. No carnivorous animal has horns; nor has any carnivorous animal hoofs like those of the reindeer.

7. Instead of hoofs, you see on the polar bear's feet, short, sharp-curved claws. These

are something like a cat's claws, only the polar bear cannot draw them in as you know a cat can.

8. But their use is the same as in the cat. The polar bear uses its claws in seizing and killing the large animals on which it feeds, just as the cat uses its claws in catching a little mouse.

LESSON V.

al'–ways	re-mem'-ber	cov'-er-ed	rea'–son
o'–cean	stretch'–ing	sea'–son	far'–ther
search	oc-curs'	car'–ry–ing	sur'–face
clum'–sy	chief'–ly	aid'–ed	hair'–y

THE POLAR BEAR. — 2.

1. Now, how did the animal get the name Polar Bear? Well, you remember how you were told in the first lesson of a part of the world near the North Pole, where the ocean is always covered with ice. That is farther north than where the reindeer lives. It is the ice that is the true home of the polar

bear, and it is for that reason that men have given it its name.

2. The border of the ice is not always found in the same place. In winter, the vast ice-fields round the North Pole cover more and more of the ocean, stretching farther to the south. Then the polar bear too comes farther south.

3. In summer, on the other hand, the ice-fields are partly melted away, so that their edge is found farther north. In that season, therefore, we must go farther north before meeting with the polar bear.

4. But neither in summer nor in winter is the polar bear to be seen often on land. When it is found there, it is, as a rule, only on the shore, though sometimes, when pressed by hunger, it will roam some miles inland, in search of land animals to kill and eat.

5. It may even happen now and then, that the polar bear is seen on land, far away from the ice-fields round the Pole. When that occurs, it is because a piece of ice carrying

one of these animals has got broken off, and has floated away on the surface of the ocean, while the bear was sleeping.

6. Now that you know where the polar bear lives, you can easily see why it must be carnivorous or flesh-eating. It can find no plants on the ice on which to feed, and it is thus driven to kill animals for its food. For this it is well fitted in many ways.

7. Though a clumsy-looking animal, it is a good runner on snow, being aided, like the reindeer, by the shape of its feet. Though so different from the feet of the reindeer in some ways, the feet of the polar bear are yet like those of the reindeer in being very large, and also in being hairy underneath. In the last lesson, you were told how these things help an animal in running on snow.

8. But the food of the polar bear is chiefly found in the water; and it is because the polar bear is so good a swimmer and diver, that it is able to live where it does.

LESSON VI.

chief'–ly	con–sists'	north'–ern	flip'–pers
broad	pad'–dles	re–main'	prey
breathe	keen'–ness	scent	col'–or
ac'–tive	ad–van'–tage	u'–su–al–ly	nat'–ur–al
val'–ue	es–teem'–ed	warmth	man'–a–ges

THE POLAR BEAR. — 3.

1. In the last lesson, you were told that the food of the polar bear is chiefly found in the water. What is this food? It consists partly of fishes, and for these the bear always has to dive and swim. But there is another kind of animal, of which the polar bear is very fond, and of which there are a great many in the northern seas.

2. This animal is the seal, which is itself a carnivorous mammal, living chiefly on fishes. It has long canines like the polar bear's; but instead of legs it has broad flippers, which serve as paddles in the water. The seal lives chiefly in the water, where it is as much at home as the fishes which it makes its prey.

3. Yet it cannot remain always under water as fishes can. No mammals can live without air to breathe; and therefore seals, when they want to breathe, must come to the surface.

4. Now, this gives a great chance to the polar bear. Where the water is covered with ice, the seal must always have a hole in the ice, to which it may come to breathe. This the polar bear knows, and it watches at one of these holes for hours together, ready to kill the seal with a single blow of the paw, as soon as it shows its head at the opening.

5. In finding these holes, the polar bear is aided by its great keenness of scent; for its sense of smell is so sharp, that it can discover seals' breathing holes by that means under the snow, just as the reindeer can discover the buried reindeer moss.

6. But seals are often caught by the polar bear on the ice, where they rest and sleep. And in capturing prey in this way, even its color is an advantage to the bear. The

young animal is as white as snow, so that it cannot easily be made out. When older, the polar bear is yellowish, but even then it is often able to come pretty close up to its prey without being noticed.

7. But when it has the chance, it always dives and swims under the ice, and then rises as near as it can get to the point where it has seen the seals. When it manages to do that, the seals cannot escape.

8. The male polar bear is always active, both in summer and winter, but the female usually passes the winter in a long sleep. It then lies buried amongst the snow, in a natural den made by the warmth of its own breath, and in the same way a long tube is kept open to the surface, so that the animal is not in danger of dying from want of air. In most of the animals that pass the winter in sleep, the breathing is then almost at a stand-still, so that hardly any air is needed.

9. The polar bear is much hunted by man. Its long soft fur is of value, and its flesh is highly esteemed as food.

LESSON VII.

wal´-rus	coast	in-stead´	ta´-per-ing
up´-per	tusks	sock´-ets	swol´-len
whisk´-ers	bris´-tles	fierce	ter´-ri-ble
crea´-tures	sev´-er-al	thou´-sand	puff´-ed

THE WALRUS. — 1.

1. THIS strange-looking animal is met with in the same part of the world as the polar bear; only the walrus is found more in the water than on the ice. while the polar bear is more on the ice than in the water. The walrus, too, lives, as a rule, nearer land than the polar bear, for it is on the coast that it finds the food it likes best.

2. The walrus is, in fact, a kind of seal. Like the seals, it has flippers instead of legs, and like them, also, it has a body thick in the middle, and tapering towards the head and tail. This form of body fits it very well for swimming through the water, just as a boat is fitted for sailing, by being made sharp at the bow and the stern.

3. The walrus is like a seal also in having the canine teeth, of which you have now heard more than once. But in the walrus these canine teeth are not like those of either the proper seals or the polar bear. They are the chief mark by which the walrus is known.

4. They are found only in the upper jaw; but there they grow to a large size, and are very large in male walruses. They form true tusks, pointing downwards, and reaching far below the lower jaw. Sometimes they are found to be as much as two feet in length, and more than seventeen pounds in weight.

5. It is chiefly these large tusks that cause the head to have such a strange form. The sockets in which they are set are holes six or seven inches round, so that the upper jaw appears greatly swollen and puffed out. On the large upper lip are whiskers like those of a cat; but these whiskers are not, properly speaking, hairs, but bristles as thick as a crow-quill.

6. Altogether the walrus is a fierce-looking creature; and when you are told that it is sometimes over twenty feet in length, and nearly a ton and a half in weight, you may be ready to suppose that it is one of the most terrible beasts of prey in the far north.

7. But this is not the case. It does, indeed, live partly on small seals and fishes; but the food it likes best is small shell-fish, like shrimps, and other creatures found on the sea-bottom near the shore, and it is also very fond of sea-weed. It is thus partly herbivorous.

8. The walrus is mostly found in large herds. At one time these herds might consist of several thousands, but now the number of walruses is not nearly so great.

LESSON VIII.

yield	ter´-ri-ble	weap´-ons	cer´-tain
fierce	de-fence´	at-tack´-ed	wad´-dles
jerks	flip´-pers	com´-rades	val´-ue
tough	har´-ness	peace´-a-ble	be-neath´
lay´-er	blub´-ber	pro-tect´-ed	pierc´-es
i´-vo-ry	har-poon	fast´-en-ed	leath´-er

THE WALRUS. — 2.

1. THOUGH as a rule a peaceable animal, the walrus can make use of its tusks as terrible weapons, when it likes. At certain seasons, there take place between the males fierce battles, in which these weapons often prove very deadly.

2. Having such a means of defence, the walrus is not afraid of the polar bear, and when it is in the water. not even of man. On land or on the ice, however, the walrus is rather a helpless creature when attacked by man. It cannot run, but only waddles and jerks along by means of its flippers.

3. The walrus is much hunted for several reasons. Its strong hide is of great value.

The leather made out of it is used for harness
and other things, for which a very strong and
tough leather is needed.

4. Then beneath the skin there is a thick
layer of fat, which is called blubber. This
is found in all mammals living in cold
seas, for all mammals in cold countries
must be protected in some way or other
against cold.

5. Those which live on land have thick
coats of fur, which serve that purpose. But
the fur of the walrus and other mammals
living in the sea is not enough; and so there
is under the skin a coat of blubber, which
helps to keep the animal warm.

6. This blubber yields an oil which is
highly prized. But the walrus is hunted
most of all for its tusks, from which is
obtained a fine kind of ivory.

7. The chase of the walrus, when carried
on at sea, is not without danger. A herd
of walruses will come boldly up to the boat
of the hunters, and raising their heads out of

the water, will try one after another to capsize the boat with a blow from their powerful tusks.

8. One of the men in the boat stands ready with a short spear, to which a rope is tied, the other end of the rope being fastened to the boat. This short spear is called a harpoon, and is made like the end of a large fishing-hook, so that when it pierces the side of a walrus it sticks.

9. But when a walrus has been struck, it darts rapidly through the water, dragging the boat after it, which is in danger of capsizing. Yet in spite of this danger, the hunters harpoon one walrus after another, until there may be as many as six tugging at the boat, while their comrades remain near lashing the water in fury with their tails. When fainting from loss of blood, the walruses are pulled to the side of the boat and killed.

LESSON IX.

film	Green'–land	breathes	bur'–ied
moist	sur'–face	wal'–rus–es	nos'–tril
whol'–ly	breath'–ing	min'–utes	moist'–ure
sail'–or	dis'–tance	whal'–ers	blub'–ber

THE WHALE. — 1.

1. THERE are many kinds of whales, and some of them are the largest of all animals. The most useful to man is that called by sailors the Right Whale, or the Greenland Whale.

2. It lives chiefly in the northern seas, and takes the name of Greenland Whale from a country near the North Pole, a country almost wholly buried under thick masses of ice, except near the sea.

3. The length of a large whale of this kind is from sixty to seventy feet. From its form, you would suppose that this animal was a fish. It has a long body, tapering towards the head and tail; the shape of the tail is like what you see in many

fishes: on the body there is no hair; moreover, it never leaves the water.

4. Yet it would not be right to call it a
fish. It is a mammal. It suckles its young
with milk; it has warm blood: and it
breathes at the surface of the water as
seals and walruses do.

5. Can any boy or girl see anything in
the picture of the whale that is not like a
fish? You may if you look at the tail. Its
shape is indeed fish-like. but see how it lies.
The flat sides of the tail are above and below,

and the tips of the tail are at the sides of the animal.

6. The tail of a fish is always placed upright, and is moved from side to side in swimming, while the whale moves its tail up and down. It is chiefly by this movement of the tail that the animal swims, but it is aided by the flippers in front, which are somewhat like the fins of a fish, but more like the paddles of a seal or a walrus.

7. The nostrils of the whale are on the top of the head, so that it has only to put the top of the head above the water, in order to breathe. It rises to breathe every eight or ten minutes as a rule, but it can remain as much as twenty minutes at a time under the water.

8. The breathing of the whale is called blowing or spouting. On rising to the surface, the whale first of all breathes out the used-up air from the lungs. Warm air coming from the lungs is always moist, as you yourself may see at any time by breath-

ing upon a slate. The slate at once is covered with a film of moisture.

9. Now, when the whale blows, this moisture in its breath is changed into a fine spray in the cold air, so that the animal seems to be spouting up water. The blowing of the whale makes a noise which can be heard at a great distance.

10. You do not need to be told now that the whale must be kept warm. You remember that the walrus has both fur and blubber for the sake of warmth. The whale has no fur, but its layer of blubber is all the thicker. Sometimes it is as much as two feet thick; and it is chiefly for the oil to be got from this blubber, that the animal is hunted and killed by men called whalers.

LESSON X.

arch	ba-leen′	es-cape′	fringe
depth	mor′-sels	har-poon′	di-vide′
throat	o′-cean	an′-i-mals	gul′-let
tongue	cu′-ri-ous	sur′-face	stif′-fen
flesh′-y	squeez′-ed	swal′-low-ed	bod′-ies

THE WHALE. — 2.

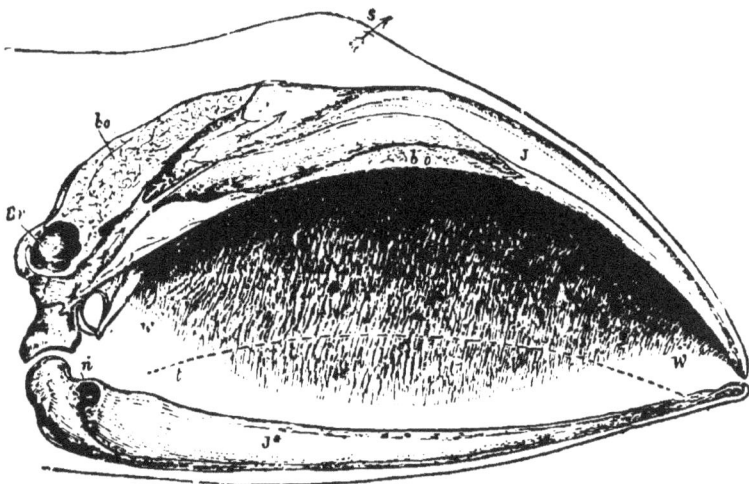

1. THE fat or blubber is not the only thing
for which the whale is killed. In the large
mouth of the whale, there is something else
of value, called whalebone, or baleen. The
use of this baleen to the whale is to help

it to catch its food, for the whale has a very
curious mode of feeding.

2. It has a very narrow throat or gullet,
so that, for so large an animal, it can swallow only small morsels of food ; besides which
it has no teeth to divide its food.   Hence it
feeds only on small animals with soft bodies,
and there are very great numbers of such
animals near the surface of the ocean.

3. The baleen, along with the tongue of
the whale, forms a sort of trap in which the
soft animals are caught.   It is made up of
a great number of horny plates, which hang
down from the roof of the mouth, and split
up at their lower end into a fringe, forming
a kind of brush.

4. The plates are of such a shape, that a
hollow is left in the middle of the mouth, and
the brush has the form of an arch.   The hollow is filled up by the large soft fleshy tongue.

5. As the whale swims rapidly through the
water with its mouth wide open, hundreds of
the small animals rush into the gaping hol-

low. The whale then presses its tongue against the baleen brush; the water is squeezed through the brush, and out of the whale's mouth, but the little animals are caught and swallowed.

6. It is the horny plates of the baleen that are most useful to man. They are used in many ways, but chiefly to stiffen some parts of women's dress.

7. In the chase of the whale, the whalers are exposed to many hardships and dangers. They must sail far away into the wintry seas of the North or South, near to the edge of the great ice-fields, of which you read in the first lesson.

8. Large masses of ice float about on the surface of the ocean, and between these a ship may get caught and crushed, as if it were a nutshell; or if it sails up a large crack in the ice-fields, it may get frozen in, so that it cannot escape.

9. To strike the animal, the whalers make use of a harpoon something like that used

in hunting walruses. But the harpoon of the whalers must have a very long line fastened to it; for as soon as the whale is struck. it dives quickly to a great depth. It stays under water as long as it can, but in the end it must rise to breathe.

10. A second time it is struck with a harpoon; and then the whale in its fury will often lash the water with its tail again and again before diving. making the water a mass of foam mingled with blood.

11. If the boat of the whalers should be too near, a single blow of the tail would be enough to throw it high in the air. Soon the poor animal is worn out by fury, pain, and loss of blood, and the whaler can then get near and kill it with a lance.

12. When dead, the whale is towed to the side of the ship, and made fast. It is now stripped of its blubber by men who stand on its body. shod with spiked boots to keep them from slipping. The blubber is at first cut off in large blocks, but on board of the

ship it is freed from useless matter, and stored in casks to be carried home to make oil. Other men at the same time take the whalebone out of the huge mouth. The oil and whalebone from a single whale has been sold for as much as three thousand dollars.

LESSON XI.

sa'-go	cit'-ies	om-niv'-o-rous	tur'-nips
length	sum'-mer	pro-vok'-ed	ber'-ries
car'-rots	an'-i-mal	al-read'-y	au'-tumn
peo'-pled	an-oth'-er	hi'-ber-nat-ing	cab'-bage

THE BROWN BEAR. — 1.

1. In this lesson and some others that follow, you are going to read about animals in countries more like our own than those round the North Pole. These countries are unlike one another in many ways; but they are all alike in having spring, summer, autumn, and winter as we have.

2. Of the animals that live in such countries, there are many that you will like to

hear about. We will begin with the Brown Bear.

3. There are many kinds of bears. You

have already been told about one that lives chiefly on the ice. That one is wholly car-nivorous; but it is the only one of all the

bears that is so. The other bears are omnivorous. which means that they live on all kinds
of food, both on the flesh of animals and on
what comes from plants.

4. Man is an omnivorous animal, for he,
too, feeds on both animal and plant food.
His beef and mutton, his fowl and fish, make
up the chief part of his animal food. His
plant food is made up not only of such things
as cabbages, pease, carrots, and turnips, but
still more of bread and fruits, rice, sago, and
so forth. All these things come from plants,
or are made from things got from plants.

5. The brown bear is not very often seen
in countries that are thickly peopled. Like
other large wild animals. it has been driven
away from lands that are covered with farms
and cities.

6. In appearance, the brown bear is very
like the polar bear, except in its color. It
has a similar shape, similar claws, and similar
teeth. Though not quite so large as the polar
bear, it is often more than seven feet in length.

7. If you saw it, you might think it a very terrible animal, both to men and to other animals. Yet it is, as a rule, harmless enough. It will not attack a man unless it be provoked, and it does not even kill weaker animals for food.

8. It prefers to live on roots and berries and other things belonging to plants. Honey and other sweet things it is very fond of. It is mostly on the approach of winter, when it is pressed by hunger, and cannot get enough of the plant-food it likes best, that it attacks animals for their flesh.

9. But when the winter has come, the bear falls into a long sleep. It is one of the hibernating animals; and both the male and the female hibernate, not merely the female, as is the case with the polar bear. Before this winter-sleep takes place, it grows very fat; and when fat enough and no longer caring to eat, it goes and lies down in its den between rocks or among the roots of trees.

LESSON XII.

droll	hun'-gry	fright'-en-ed	gam'-bols
en-joy'	qui'-et-ly	mis-be-have'	moth'-er
for'-est	climb'-er	a-mus'-ing	per-form'
ter'-ror	sleep'-i-ly	dis'-tance	shag'-gy
wom'-an	awk'-ward	straw'-ber-ries	e-nough'

THE BROWN BEAR. — 2.

1. THOUGH very strong, the bear is for the most part a gentle animal, unless when old or hungry. Even a wild bear has been known

to play with children like a big dog.

2. A story is told of a woman in Russia, where bears are not uncommon, losing two young children, and being much frightened at finding them playing with a bear and the bear with them.

3. One of the children was feeding the

great shaggy beast with strawberries. which it seemed greatly to enjoy; the other was riding on its back. quite happy, and never thinking that the horse it had mounted could do it any harm. Nor did it do so; for in its gambols it behaved as gently as a large, good-natured dog would have done.

4. When the bear saw the mother coming near in terror, it turned and walked quietly away into the forest.

5. Clumsy as it looks, the bear is not so awkward in its movements as we might fancy. It is a good runner, a good climber, and a good swimmer. When caught young, it can be trained very easily to dance and to perform amusing tricks; and when kindly treated, it becomes very fond of its master.

6. Even when left to themselves young bears, or bear cubs as they are called, are very amusing. In some places they are kept in large pits, and people often come to watch their droll ways, as the plump little animals play with one another, and enjoy themselves in their own way.

7. The mother will sleepily watch them meanwhile, but will sometimes rouse herself to give a smart cuff with her paw to any of her children that misbehave.

8. During their sports, one will often see the young bears rise on their hind-feet, and use their fore-legs as if they were arms. Their hind-feet have flat soles like the feet of men, and hence bears can stand upright almost as firm as a man can.

9. When full-grown, bears will often get up and stand or run for a short distance in the same way. When set upon by hunters, a bear will often turn upon them; and then it always gets up on its hind-legs, to make a rush at the man standing nearest.

10. It will then try to fell him with a single blow from one of his strong fore-paws; and if it fails to do this, it will seize him round the body with both its paws, and hug or squeeze him to death. It is in this manner, too, that the bear kills animals for food.

LESSON XIII.

i-de′-a	plen′-ti-ful	al-though′	shag′-gy
broad	shoul′-ders	re-quir′-ed	red′-dish
bi′-sons	buf′-fa-loes	Eu′-rope	ap-pears′
trav′-el	cloth′-ing	else′-where	re′-al-ly
is′-lands	A-mer′-i-ca	nu′-mer-ous	prai′-ries

THE BISON.—1.

1. THE animal called by this name is like a very large ox or cow; but if you look at a picture of it, you will see at a glance one thing by which you may always know it from an ox. It is very high and broad at the shoulders; and it appears to be even higher there than it really is, because it has so much long shaggy hair in that part.

2. The back slopes down from this hump to the tail. and there is another slope still steeper from the hump to the head, which is much lower than the shoulders.

3. The bison has horns like a bull; but the horns are always short and curved, and placed very far apart. and at the sides of the head.

4. It was in our own country that bisons were most plentiful. They are often called buffaloes. although that name is given elsewhere to animals of another kind.

5. They lived in herds. made up of count-

less numbers of animals. Nothing now can give us any idea of the size of the vast herds which used to roam over the plains. The bison is now greatly reduced in numbers, and before long will be extinct.

6. Like the cow, they feed on grass and herbs; and in the western plateau of North America there are vast plains called prairies, covered with grass, on which you can travel for thousands of miles and see little but grass on all sides. There are no fields, no hedges, no trees except here and there by the sides of rivers, and in many parts no hills or mountains.

7. It is these grassy plains that were the home of the bison, and until recently no other large animal in the world was found in such countless herds. But when the herds were larger the only human beings on these plains were Indians, who lived by hunting the bison.

8. These Indians are people who have a brownish or reddish skin, and sleep in rude tents. From the bison they used to obtain almost all that they required, just as the Lapps do from the reindeer.

9. At that time the bisons were far more numerous than the people in North America.

It is in fact said, that only one vast herd roamed over all the thousands and thousands of miles in the prairies.

LESSON XIV.

guide	quick'–ly	ter'–ri–fi–ed	tal'–low
scour	trod'–den	head'–long	use'–ful
pis'–tol	won'–der	crowd'–ing	stu'–pid
pan'–ic	stum'–ble	un–spar'–ing	mon'–ey

THE BISON.— 2.

1. AT the end of the last lesson you were told about the great numbers of bisons that once lived in North America. It is different now that people from other countries, in Europe, came over to America, and made fields to grow corn and fruit in many parts of the prairies, and laid railways across them.

2. The people of Europe have also brought with them horses and guns, which, before they came, were not known in America. Since they came, therefore, the bisons have been killed in much greater numbers. The

vast herd has been broken up into smaller ones; and in no long time the bison will be quite destroyed.

3. You may wonder why the bisons are killed in that unsparing manner. The reason is, that so many parts of the animal are useful to man. By selling the hides, the horns, the hoofs, the flesh, and the tallow, a hunter can make much money, and therefore he always kills as many as he can.

4. The hunting of the bison is very easy. Large as the animal is, it is very timid and also very stupid. When it catches sight of a hunter it at once takes to flight. Very often a whole herd will take fright at some simple thing. and dash off in a panic.

5. Crowding together, they scour away over the plains so swiftly, that no horse can keep up with them for very long.

6. Sometimes the hunter, mounted on a good horse, will dash into the very thick of such a terrified herd, and kill the animals with his gun or pistol right and left. He

knows that there is no danger of the bison
itself setting upon him.

7. Yet he knows, too, that when he chases
them in this way he takes his life in his
hands; for he cannot see where he is going,

he cannot guide his horse, and at any time
his horse may stumble in a hole and get
overthrown. When that happens, the hunter
is apt to get trodden to death by the bisons
galloping over him in their headlong flight.

LESSON XV.

jaws	straight	stom´-ach	ought
chew	mere´-ly	di-gest´-ed	out´-er
hoofs	di-vid´-ed	care´-ful-ly	ly´-ing
mode	swal´-lows	be-gin´-ning	pel´-let

THE BISON. — 3.

1. AT the beginning of the first lesson on the bison, you were told that this animal is like the ox or cow. The likeness spoken of there, was only the likeness of the outer form of the two animals.

2. But you ought to know also that these animals, and many others besides, are like one another in their mode of feeding.

3. When a cow grazes, or walks about in the fields eating the grass, it does not chew the grass, but swallows it at once into a stomach, where it is only stored up for a time, to be used again.

4. But grass, like all food, must be digested ; which means, that it must be changed in such a way that it can form

part of the flesh of the animal. This change takes place chiefly in the stomach; and in most animals. as in man. there is only one stomach.

5. The cow and the bison. however. have more than one stomach; and so too have most other animals which have the feet divided into two hoofs. All these animals pass their food first of all. without chewing, into a stomach. in which it is merely stored up.

6. Now. if you watch a cow lying down in a field. you may see how it uses this store of food afterwards. You may see it working its jaws carefully from side to side, chewing and chewing for a long time without biting the grass, and without taking up any food by means of its mouth.

7. But if you watch closely. you will see where this food comes from. You will see the cow from time to time raise its head, and make the front of its neck stiff and straight, and then too. you can see a swelling passing up the front of the neck towards the mouth.

8. This swelling is caused by a pellet or little ball of grass passing up from the store already formed into the mouth. It is this pellet that is chewed so carefully. There is a name for such a pellet of food drawn up from the cow's stomach. It is called a *cud*, and animals that feed in this way are said to chew the cud.

9. When the chewing is done, the food is swallowed into another stomach, where it is digested. The cow and the bison have four stomachs. In two of these the food lies before it passes back to the mouth, and into the two others it passes after the chewing of the cud.

LESSON XVI.

scales	A-mer′-i-ca	plen′-ti-ful	chis′-els
group	gnaw′-ing	prai′-ries	for′-ests
bea′-ver	pop′-lars	cloth′-ed	rab′-bit
far′-ther	ro′-dents	birch′-es	edg′-es
wil′-lows	cu′-ri-ous	gnaw′-ers	lil′-ies

THE BEAVER. — 1.

1. THE Beaver, like the bison, is chiefly found in North America. Look on the globe for the northwest of America. That part is one in which there is a great number of lakes, or sheets of water, and a great number of rivers, both large and small. There, too, are vast forests. It is in that part that the beaver is most plentiful. The prairies with the bison lie farther south.

2. The beaver is one of a large group of animals which are all very like one another in the shape of their teeth. To this group the rabbit and the mouse also belong; and if you get the chance of looking at a mouse's head, you may see what the teeth are like.

3. In the front of the mouth you will see only four teeth, two in the upper jaw and two in the lower, and these teeth have all sharp cutting edges like chisels. Behind these teeth, there is a wide gap in the mouth without any teeth; and then come a number of other teeth set close together, and all flat at the top.

4. These back teeth are used to grind the food. but the sharp front teeth are used in gnawing; and hence the animals making up this group are called rodents, which means gnawers.

5. The beaver is one of the largest of the rodents, and is much larger than a rabbit. It is about two feet in length, and has a broad. flat tail about one foot long. The body is clothed with a fine brown fur, which is highly prized; but the tail is naked, or, rather, covered with scales instead of hair.

6. Of all the rodents the beaver is the most wonderful. It is by far the most knowing; and its ways of life, where beavers

are numerous, are very curious. It lives mainly, but not wholly. in the water. and always has its home beside streams or lakes.

7. Its food is made up of the roots of water-lilies, and other plants that grow in the water, and the bark and young wood

of trees, such as birches, willows, and poplars, which grow by the sides of lakes and rivers.

8. Where there are not many beavers, their home is merely a hole made by burrowing in the earth, by the side of some stream or lake. The entrance is a tunnel opening under water, and there is always a small opening above to admit air

9. The chamber to which the tunnel leads is hollowed out above the highest level that the water ever reaches; for if it were not, the water would rise into it and drown the beavers. The inside of the chamber is lined with withered leaves and other things to keep it warm.

LESSON XVII.

bowl	plen′-ti-ful	cham′-ber	skil′-ful
built	won′-der-ful	sev′-er-al	gnaws
twines	plas′-ter-ed	en′-trance	trow′-el
rap′-id	nar′-row-er	reg′-u-lar	straight
sol′-id	car-niv′-o-rous	com-posed′	eigh′-teen

THE BEAVER. — 2.

1. WHERE beavers are very plentiful, their homes are quite different from those you read of in the last lesson. The dwellings they make for themselves are as wonderful as those which any other animals are known to make.

2. The beaver lodges, as they are called, are round; they are built in the water, and rise several feet above the surface. The part above the water is shaped somewhat like a haycock, or a bowl turned upside down.

3. The things which the beaver makes use of in building these lodges, are the branches of trees, stones, mud, and mosses. The branches it gnaws off with its strong gnaw-

ing teeth ; and in laying them it twines the twigs of different branches together, so as to keep each other firm, and the branches laid under water are kept down by stones placed on the top of them. The mosses and mud are used to fill up the spaces.

4. All the outside of the lodge above the water is carefully plastered over with mud, and in laying on this covering it is very likely that the beaver makes use of its scaly tail as a trowel. The walls of these lodges are at least three or four feet thick, and are hence very strong.

5. The lodges are, indeed, regular forts, into which the carnivorous animals that make the beavers their prey cannot break their way. In winter, when the walls are made solid by the frost, they are even stronger than in summer.

6. Inside the walls of the lodge is a chamber about seven feet across and three feet high. In each chamber may live five or six beavers. Their beds, composed of dry

leaves and grass, are ranged round the walls. In such lodges the beavers also keep stores of food for the winter.

7. The entrance to the lodges is always below the surface of the water, and in streams that become very low in summer, the beaver has a skilful way of keeping the water always high enough. It builds dams across the stream from bank to bank.

8. Sometimes such dams are three hundred feet in length. Since they must be made very strong to stand the force of the water, they are very thick. At the bottom their thickness is ten or twelve feet; but they get narrower towards the top, where their thickness is only about two feet.

9. The logs used in building them are sometimes eighteen inches or even two feet thick; but however large the logs are, a countless number of them must be used in building such large dams.

10. Where the flow of a river is gentle, the dams are built straight across; but where

it is more rapid, they are curved up the stream so as to stand the force of the water. In a like manner, a builder places the bricks over a doorway or window in the form of an arch, to bear up the wall put on it.

LESSON XVIII.

oil′-y	no′-ticed	sur′-face	pounce
dif′fer	u -nit′-ed	feath′-ers	hab′-its
grubs	chick′-ens	bis′-cuit	pig′-eon
known	swoop′-ing	sud′-den	suck′-le
plough	keen′-eyed	cov′-er-ing	pad′-dles

THE GULL.

1. This is the first bird that you have read about; and before you are told anything about the gull itself, it will be well for you to know something of how all birds differ from the mammals, of which you have been reading.

2. You hardly need to be told that birds do not suckle their young. They lay eggs; and when the young birds come out of the

eggs, they are sometimes able to pick up their own food at once, as chickens can.

3. Birds have no hair, but they have feathers to cover them; and this covering serves to keep them warm, just as the hair or fur of mammals does.

4. The common sea-gull is a white sea-bird, larger than a pigeon, and is known to most boys and girls who live by the seaside, or near the mouths of rivers.

5. By watching it from the shore or from the deck of a steamer, they can learn something of its habits. They can see how keen-eyed it is. They may have noticed it swooping down with a sudden dart, to pick up a bit of bread or biscuit, which it will not fail to see, even amidst the foam of a steamer's wake.

6. Just so they may see it pounce down again and again to the surface of the water, trying to catch small fishes. But often it may be seen to fly back again without anything in its mouth; for, though keen-eyed

and quick on the wing, it is not a good diver, and the fish after which it darts can easily escape by sinking in the water.

7. Indeed, if living fishes formed the chief food of the gull, that bird would not thrive very well. The gull, however, lives chiefly on what it picks up on the seashore, or finds floating on the waves. Often it may even be seen a good way inland, following the plough, ready to pick up worms and grubs when they are turned up with the earth.

8. Yet it is a true sea-bird; and there are two things that you ought to take note of, as fitting it for a sea life. Like all water-birds, it has its feathers covered with an oily coating, which keeps them from getting wet.

9. If you have ever seen a duck swimming in a pond, you may have noticed that when it dives, and rises with water on its wings, the water does not wet the feathers, but lies in round drops, which easily roll off when the wings are shaken.

10. The other thing that you should take note of, is the shape of the feet. The toes in front are united by a web, in such a way that the feet can act as paddles. By means of these the gull can swim when it likes.

LESSON XIX.

eaves	sel'-dom	dis'-tant	au'-tumn
tough	pas'-sage	swal'-low	cer'-tain
hab'-its	chim'-ney	un-used'	fork'-ed
rap'-id	in'-mates	quit'-ted	close'-ly
a-lights'	mi'-gra-tory	shel'-ter-ed	re-pairs'

SWALLOWS.

1. The Swallows are the best known of all the birds of passage or migratory birds, which means, birds that visit us only at certain seasons, and spend the rest of the year in distant countries.

2. Almost every boy and girl can tell a swallow by its flight. We may not be able to see it very often close at hand, for it seldom alights on the ground, or rests any-

where long enough for us to look at it closely; but we can tell it by the shape of its wings, and by its way of darting about in the air, now high, now low, with many rapid turns.

3. The common or chimney swallow is also known at once, by its long forked tail, the outer feathers of which are more than half as long as the whole body.

4. But besides this common swallow, there are other birds which are often called by the same name, and which have the same habits.

5. Of these, the birds called swifts are most unlike the common swallow, and are easily known by their color and the shape of their wings. While the swallows are black and white, and have wings that are not very long, the swifts are almost all black, and have wings both long and narrow.

6. The swallow's nest is as well known as the bird itself. It is built of small round lumps of clay, held together by straw and tough, dry grass. Inside it is warmly lined

with feathers. The nest is always built in some place sheltered from wind and rain; sometimes in an unused chimney, sometimes under the eaves of a house, or in the upper corner of a window.

7. To this nest the swallow returns year by year, after its wonderful migration. It is absent from the Northern States during all the winter months, and during that time it lives in countries many hundreds of miles away.

8. Yet in the spring-time it finds its way back to the same part of the country which it had left in autumn; and when it has once built a nest, it does not build a new one, but merely repairs the very nest which it quitted.

9. A swallow may thus get well known to the inmates of a house where it has built its nest. One has been known to come for about ten years to a nest in the porch of the same house; and while it paid no heed to the passing out and in of the people belonging to the house, it would utter a cry of warning when any stranger entered.

10. This habit of migrating shows a wonderful instinct, and you may well ask the reason of it. One reason no doubt is, that the bird cannot stand the winter cold; but there is another reason besides. The swallow feeds on flying insects, which in the north of our country are very scarce in winter, so that the bird flies away to a warmer climate, where they are to be found. In a later lesson you will learn where such countries are; countries that have a summer all the year round.

LESSON XX.

in´-sects	chil´-dren	gath´-er-ed	flut´-ter
bod´-ies	cu´-ri-ous	weath´-er	cloth-ed
hid´-den	to-geth´-er	stretch-ed	coun´-tries
sel´-dom	as-sem´-ble	ap-proach´	jerk´-y

BATS.

1. CHILDREN who live in the country know these curious animals very well. They can see them flying about at dusk. and though it may be too dark to see what the bats are

like, yet they can easily tell them from birds,

by their manner of flying, which is always
very jerky. Their wings flutter more than

those of birds do, and the bats are seldom seen to alight.

2. They are very curious animals indeed. They are flying mammals, and they are the only mammals that do fly. If you could catch one, you would see that they have their bodies clothed, not with feathers like birds, but with hair like other land mammals. Their wings, too, are very unlike those of birds. They are nothing but a very thin skin, which can be stretched out, by means of long, thin bones like very long fingers.

3. It is only by night or in the dusk that these strange mammals fly about. In our part of the world, their food consists of insects, which they catch flying as swallows do. By day they sleep in holes in trees or in other places where they can remain hidden.

4. Bats, like swallows, are not to be seen in winter. Well, you know that bats feed on the same kind of food as these birds do, and you may perhaps think that the bats

also fly away in winter. to countries where there is summer all the year round. But no, that is not what bats do. They are able to pass the winter in a very strange manner.

5. During that season they sleep, and thus are able to do without food till the warmer weather returns.

6. On the approach of winter, they assemble in great numbers in some cave or cleft in the rocks, in large hollow trees, in barns. in sheds or old buildings ; and they all hang by the claws of their hind feet. close together, head downwards.

7. Like the bear and other animals that pass the winter in this way, the bats are warm and fat when their winter sleep begins. but cold and lean when it ends.

8. In our part of the world the bats are all small; but in those warm countries, of which you have heard, with summer all the year round. there are very large bats which live on fruits. and do not have a long sleep.

9. They are hence called fruit bats. Some

of them are so large. that their wings, when spread out. are nearly five feet from tip to tip.

10. Like our own bats, they fly about by night, and by day they hang head downwards from the branches of trees or the roofs of caves.

LESSON XXI.

bursts	dif′-fer-ent	creat′-ure	emp′-ty
jel′-ly	bus′-i-ness	mul′-ber-ry	re′-al-ly
thread	sev′-er-al	cab′-bage	in′-sect
cir′-cle	wing′-ed	squeez′-ed	dur′-ing
moths	sup-plies′	na′-tives	use′-ful
strands	pur′-pose	cat′-er-pil-lar	co-coon′

THE SILKWORM.

1. Though called a worm, this creature is really an insect, and in one state it has legs and wings like a butterfly. But a great number of insects are quite different at one time of their life from what they are at another.

2. When they are first hatched from the

egg, they are creeping, worm-like creatures, like caterpillars which you so often see on cabbage - leaves. Afterwards they become quite still for some days, and you would take them to be dead.

3. But during that time, great changes

are going on within t h e i r o u t e r covering; a n d at last a winged insect like a b u t t e r f l y b u r s t s out, and leaves the case empty.

4. The creature that supplies us with the silk of which dresses are made is called a worm, because it lives longest in the state in which it is like a worm, and also because it is then useful to man.

5. It remains a caterpillar for about eight weeks. At the end of that time, it begins to spin around itself a small ball of silk called a cocoon, and it goes on spinning for about five days.

6. If you opened a silkworm while it was spinning its cocoon. you would not see anything like silk inside it. You would see two masses like jelly, and it is this that becomes changed into silk.

7. In the mouth of the silkworm are two very small holes, by which the jelly is squeezed out; and as it passes into the air. it hardens into a very fine thread.

8. Though there are two holes, there is only one thread, because the two streams of jelly get joined into one as soon as they pass out. Yet with a glass which makes very small things look larger, you can see that there are really two strands, even in this very fine thread.

9. To spin its cocoon out of this thread, the silkworm keeps slowly turning its head

in a circle. While this is going on, a slight noise may be heard; and when the noise stops, those who watch the silkworms know that the cocoon is finished.

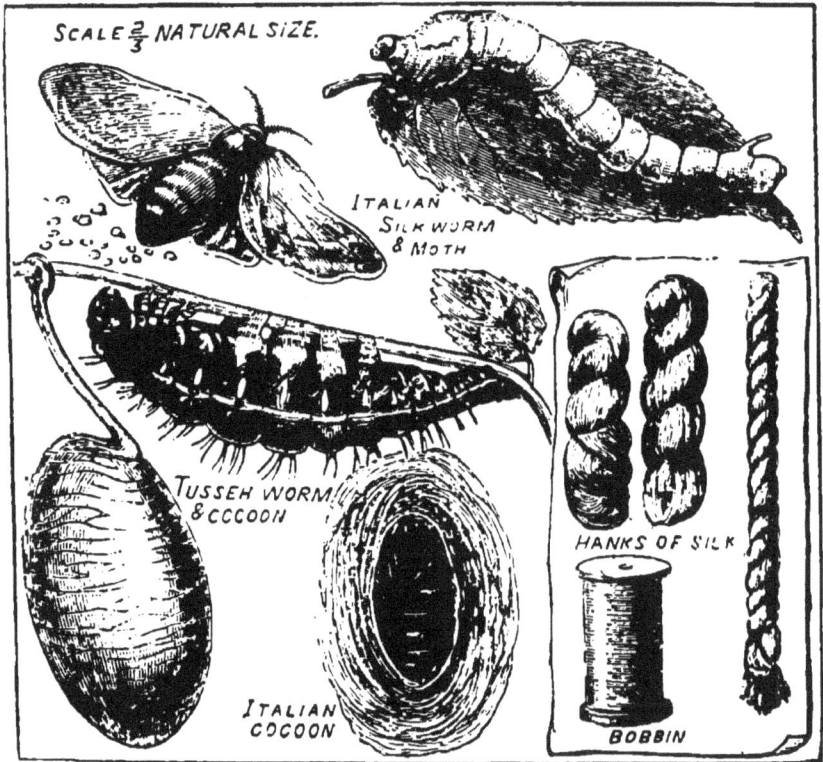

10. If this cocoon were left to itself. the worm would in the end change into a moth inside. and the moth would eat its way through the silk. As soon, therefore, as

the noise stops, the cocoons are put into an oven to kill the worm. Only a few cocoons are left to be eaten through by the winged moths, in order that these may lay eggs, to bring forth new silkworms for the next year.

11. So fine is the thread which the silkworm spins, that three hundred yards or more are wound off from a single cocoon, and the threads of several cocoons must be spun together to make a thread fit for being used in weaving.

12. Silkworms are natives of warmer countries than ours; and where they are reared in colder countries than their own, stoves must sometimes be used, to give them the warmth which they need.

13. In our country it is not easy to rear them, but in the south of Europe the rearing of silkworms is a great business. The best food for them is the leaves of a tree called the mulberry, which is grown for the purpose.

LESSON XXII.

in′–ches	ev′–en–ing	ea′–ger–ly	tend′–ed
sin′–gle	dread′–ed	chief′–ly	mass′–es
plagues	pun′–ish–ed	trav′–el–lers	re–fus′–ed
lo′–cust	re–mem′–ber	men′–tion–ed	ac–count′

THE LOCUST.

1. Like the silkworm, the Locust is an insect found in warm countries; but, instead of being reared and carefully tended by man, for the sake of what is got from it, it is almost everywhere dreaded on account of what it destroys.

2. It is a flying insect, and lives chiefly on green herbs. Though large for an insect, being about two and a half inches in length, a single locust cannot, of course, destroy much; but these insects often appear in countless hosts, and this it is that makes them so terrible.

3. Being found in Palestine and the countries round, they are often mentioned in the Bible. A visit of an army of locusts was one

of the plagues with which the people of Egypt were punished when they refused to let the Israelites go.

4. In the story of that plague, it is said the locusts " covered the face of the whole earth, so that the land was darkened; and they did eat every herb of the land, and all the fruit of the trees which the hail had left; and there remained not any green thing in the trees, or in the herbs of the field, through all the land of Egypt."

5. Such visits of locusts are also told of by travellers in our own times. The swarms of locusts borne along by the wind filled the air; they form thick masses, dark as a thunder-cloud. For more than an hour such an insect host has been seen to fly past, and the host may keep flying on as long as the sun shines; but when the evening arrives, the insects always fold their wings and settle down.

6. Then on all sides there is nothing seen but locusts. The plain is covered with them.

Every green thing is hidden by their vast numbers. The sound of their jaws as they bite off the herbs can be heard at a great distance.

7. Yet the locust is welcomed in some places. In deserts, where no crops are grown, and there are no pastures on which herbs can be reared, locusts sometimes supply a food that is much enjoyed.

8. In some parts smoky fires are made when locusts appear, that the locusts may be killed with the smoke. The dead locusts are eagerly eaten both by man and beast. Wherever honey can be got, it is eaten along with the locusts; and this, you may remember, was the food of John the Baptist in the wilderness.

LESSON XXIII.

juice	but'–ter–fly	cap'–tur–ed	ac'–tive
chaff	hol'–low–ed	thresh'–ing	bod'–ies
vis'–it	wel'–com–ed	to–geth'–er	warmth
hab'–its	won'–der–ful	cham'–bers	won'–der

ANTS.

1. IF you visit the country, you may chance to see many active little brown insects, with bodies shaped like what is shown in the picture. These are ants, and they are in many ways the most wonderful of all insects.

2. They live together in great numbers, and no other animals behave in many things so much like man as they do. There are many kinds of them, and they do not all have the same habits ; but you will wonder greatly when you are told of the habits of some of them.

3. They live in nests, which, as a rule, are made in damp earth, the earth being partly hollowed out, partly raised up into a mound.

Inside of the nest, are large numbers of
chambers joined by tunnels.

4. Within the chambers are kept the eggs
of the ants, and the young insects before
they are able to look after themselves. For

young ants, like young children, are quite
helpless, and must be fed and taken care of
in every way.

5. They are always kept clean. When
the sun shines, they are carried from the

inner part of the nest to the part near the top to enjoy the warmth. When it becomes cold, or when rain seems likely to fall, they are carried back again.

6. When full-grown, most of the ants are not winged insects, like the silkworm, moth, or the butterfly. Only a few are winged, and even these do not keep their wings long. Those which never have wings are called workers, and do all the work in the nest, as well as in getting food.

7. But the strangest things about the ant have still to be told. Did you ever hear of animals that kept cows of their own ? That is what some ants do. The cow of the ant is a smaller insect, which yields a sweet juice of which the ant is very fond. To get its cow to give out this juice, the ant strokes and pats it gently.

8. Some ants even have fields of their own, like the fields in which men grow corn. They mark out plots round which they build walls of earth. In these they

do not sow seeds themselves, but the fields are made in places where there grows a kind of grass, on the seeds of which the ants feed ; and the ants do not allow anything to grow in their fields except this grass.

9. All other plants they weed out, and throw over the walls, so that this grass may grow up in plenty. When it is ripe, they free the seed from the chaff, as men do in threshing corn, and the seed they store up for food.

19. Some ants, again, keep slaves to do all the work for them, the slaves being ants of a weaker kind, which are captured in their nests. But it would take many lessons to tell all the wonderful things that are done by ants, which are the most knowing of all insects.

LESSON XXIV.

suck'-le	cu'-ri-ous	sur'-face	chief'-ly
whol'-ly	spawn'-ing	col'-or-ed	sal'-mon
im-pure'	breath'-ing	sev'-er-al	hatch-ed

THE SALMON. — 1

1. In some of the first lessons, you have read of several animals that live in the water, either wholly or chiefly; but not one of these could rightly be called a fish. The Salmon, however, is a true fish. Let us see, then, how it differs from other animals that live in the water, but which are not fish, such as whales.

2. First, you remember what was said about the breathing of whales and other sea-mammals. You were told that they must come to the surface to breathe. But the salmon does not need to do this.

3. Just like man, a whale or a seal or any other mammal has lungs within the chest; and it is the lungs that fit them all for breathing air. All the blood, after being

made impure and dark-colored in passing through the body, comes to the lungs to be made pure and bright again by means of the air which is breathed in through the nose.

4. The salmon, however, has no lungs, but breathes in the water by means of *gills*, as they are called. In any dead fish, you have only to open the slit at the side of the body just behind the head, in order to see these gills, which have the form of red bands. Now, in fishes, all the blood comes to be made pure in the gills, by means of something in the water that passes over them.

5. Then, again, a salmon does not suckle its young. It lays eggs as a hen does, and it leaves the eggs to be hatched by the heat of the sun. The eggs are not, however, laid one by one as those of hens are. They are small, and are laid in very great numbers. The laying of the eggs is called *spawning;* and as it is curious in many

ways, you will be told more about it in the next lesson.

LESSON XXV.

spawn	sud'–den–ly	hatch–ed	straight
rap'–id	qual'–i–ties	shal'–low	es–cape'
grilse	up'–wards	mi'–grates	grav'–el
ac'–tive	creat'–ures	in'–sects	ang'–ler
na'–tive	mi'–gra–tory	suc–ceeds'	con–sist'

THE SALMON. — 2.

1. THE salmon always spawns, or lays its eggs, in rivers. It does not always live in rivers. It is a migratory fish, as the swallow is a migratory bird. It migrates from rivers to the sea, from fresh water to salt water, and then again from salt water to fresh.

2. In order to spawn, the salmon comes from the sea to the rivers ; and it swims up the rivers till it comes to a spawning-bed that it likes. In doing so, it sometimes meets with waterfalls ; but these do not keep it back.

3. The salmon is one of the most active of all fishes, and has very great powers of leaping. When it comes to a waterfall, it bends its body so that the head and tail come nearly together; and then, by suddenly making itself straight again, it takes a great leap out of the water, trying to reach the top of the fall. If it does so, it swims quickly upwards in spite of the rapid rush of the water. It may fail; but when it fails, it tries again and again till it succeeds.

4. At last it reaches the spawning-bed. This consists of a bed of fine gravel in a shallow part of the river, where it can be warmed by the rays of the sun. This is one reason why the salmon comes to the river to spawn. At the bottom of the sea, the eggs would not get heat enough from the sun to become hatched.

5. The eggs are not merely laid on the spawning-bed; but they are buried under the gravel, the fish making use of its tail to spread the gravel over the eggs. Many

of these eggs are eaten by other fishes and by insects; but so great is their number, that many others escape and get safely hatched.

6. When newly hatched, the salmon are tiny creatures only about half an inch in length. You can almost see through them. They are not even able to feed themselves; but part of the yolk of the egg remains fixed to the under side of their bodies, and this gets used up as food.

7. The young salmon stay sometimes for one year, and sometimes for two years, in the river in which they are hatched; but in the end they swim down to the sea. There they find plenty of food, and grow very quickly. A few months afterwards, they return to their native river, and they are now called *grilse*. They are not yet perfect salmon.

8. But soon they return to the sea again, and grow larger still; and then they gain all the qualities that make them so highly prized as food, and make them such sport for the angler.

LESSON XXVI.

is'-land	clev'-er-ly	fore'-paws	rea'-son
length	creat'-ure	fright'-en	fif'-teen
gal'-lop	cu'-ri-ous	hur'-ry-ing	al'-most
des'-ert	Aus-tra'-lia	kan'-ga-roo	con-tain'

THE KANGAROO.

1. On a map of the world, you will see to the southeast of Asia a large island called Australia. On a globe, this island will be found on the under side, because it lies nearer the South than the North Pole. It is thus a long way off from us. The fastest steamers that sail from this country cannot reach it in less than six weeks or more.

2. This island is about as large as the country in which we live, and on account of its great size is sometimes regarded as a

continent; yet it does not contain as many people as the State of Ohio, and the people that are on it live mostly in the parts near the sea. The reason of that is that almost all the inner parts are hot, sandy deserts, where rain seldom falls, and no crops can be grown.

3. The animals found on this great island are very unlike those which are met with in any other part of the world. The strangest of all, perhaps, is the kangaroo. This animal is curious in many ways, but the most curious thing about it is the way in which it rears its young.

4. It is very large. When it sits on its hind-legs and tail, as it is in the habit of doing, it is in many cases taller than the tallest man. Yet its young when new-born are only about an inch in length.

5. These little creatures are quite helpless, and almost shapeless at first, and cannot move about. And the strange thing about their bringing up, is what the mother

does with them until they are large enough
to feed on the grass, as she herself does. In
front of her body, the mother has a pouch

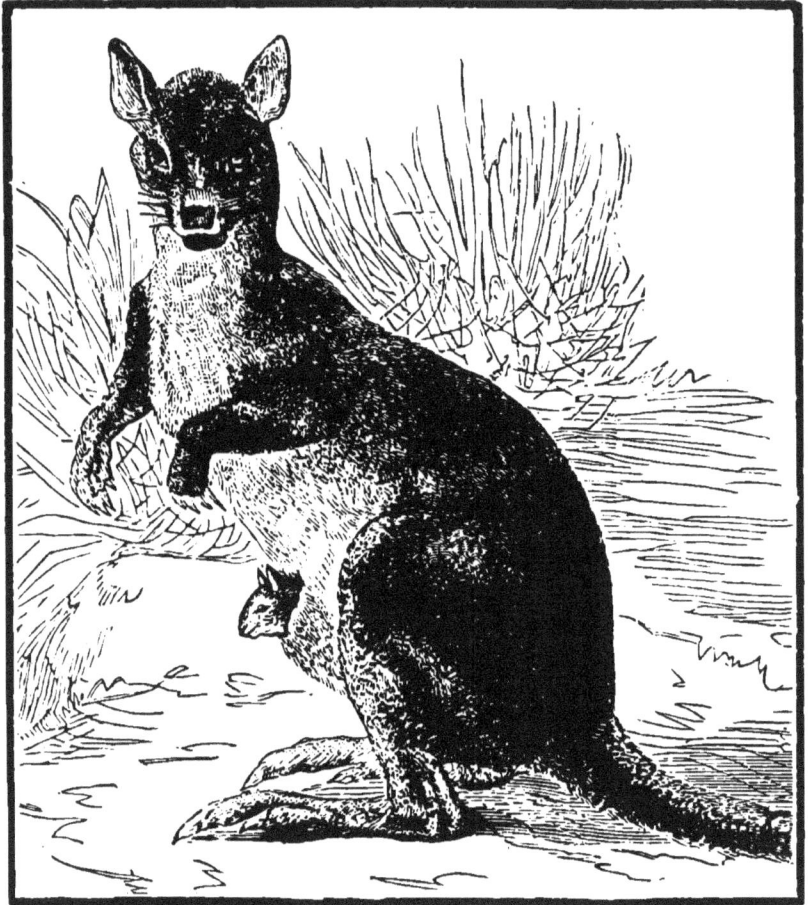

made by a fold of the skin; and into this
she puts her young when new-born, and
feeds them with her own milk.

6. About eight months pass before the young kangaroo is able to leave the pouch, and for some time after that it will often spring back again very cleverly when anything takes place to frighten it.

7. This animal is often hunted in Australia, both for the sake of its flesh and for the sport, for it can sometimes tire out the best horses. It does not run, and it does not gallop, but it takes long bounds.

8. Using its two long hind-legs and its strong tail to jerk itself upwards and forwards, it leaps about five yards at a time; and it can go on taking such leaps for fifteen or twenty miles. A strong man at a run could take only one leap of the same length.

9. In hurrying along in this way, the kangaroo never touches the ground with its short fore-paws. These paws, indeed, are hardly used as feet. They are used more as hands to pluck the long grasses on which the kangaroo often feeds, and to carry them to its mouth.

LESSON XXVII.

broad	tor′–rents	im–mense′	ex–cept′
halves	e–qua′–tor	monk′–eys	cool′–est
trop′–ic	weath′–er	an′–i–mal	for′–est
height	dif′–fer–ent	gam′–bol	tan′–gled
boughs	chat′–ter–ing	di–vid′–ed	os′–trich
les′–son	ever–last′–ing	sur–rounds′	per–haps′

THE TROPICAL OR HOT COUNTRIES.

1. In the first lesson, you read of the countries in the far north, where the winters are so long, so dark, and so cold. Now you are going to read about animals in countries of a very different kind; and before you read about the animals, you ought to know something of the countries in which they live.

2. If you again look at a globe, you will find a broad belt which surrounds its widest part, lying between two dotted lines called the tropics. It is divided into halves by a line known as the equator. Within this belt lie the countries of which we are speaking, and which are called the Tropical or Hot Countries.

3. In these countries, there are no winters at all. Snow and ice are never to be seen, except on the tops of a few very high mountains; and the people spend their whole lives in the midst of everlasting summer. Even in the coolest of these countries, the weather is very warm nearly all the year, and is never what we should think cold, except perhaps at night.

4. For most of the year, the weather is also very dry and bright; but when the rain does come, it pours down in such torrents, that some rivers rise as much as twenty feet in a single week. Moreover, the days and nights are always of nearly equal length. The tropical countries are therefore very bright, sunny places. In truth, they would be charming lands to dwell in, if they were not so very hot.

5. Among the wonderful sights in these countries are the great forests, some of which are so large that they would much more than cover the whole of the State in which

we live.   Many of the trees in them are also
of immense size, their height being twice that
of the highest of our trees, or six or seven
times that of a good house of two stories.

6. Beneath these giant trees plants grow
in countless numbers, and twine in clusters
round the trunks and boughs.   Indeed, the
network of plants is so thick and tangled
that the forests can be entered only by
narrow paths, cut through them with great
labor.

7. The largest land animals in the world
are found in the hot countries.   Elephants
force their way through the forests to feed
under their deep shade; monkeys gambol
about among the boughs, chattering away
at one another; while beautiful birds, with
gayer feathers than any we see near home,
build many sorts of curious nests.

8. Where trees are not so common the
lion makes its den among the rocks, and
countless herds of two-hoofed animals roam
over wide, grassy plains.   Along with these

latter may sometimes be seen the ostrich, that strange long-necked bird, which cannot fly, but which runs as fast as a railway train, and which lays its large eggs in holes in the sand.

LESSON XXVIII.

guess	si'-lent-ly	ca'-nines	en'-e-my
ed'-ges	warn'-ing	fear'-ful	close'-ly
sheath	al-read'-y	whisk'-er	e-nough'
pu'-pil	car-niv'-o-rous	vel'-vet-y	mus'-cles

THE LION.—1.

1. The Lion is a large cat. Boys and girls who have seen a wild-beast show may not think that the two animals are very like one another, but yet there are many things about them in which you can see that they are like when you look closely enough.

2. Look at the teeth when the lion opens its mouth. There are rows of small teeth in the front of the mouth; but at the sides in each jaw are two large sharp-pointed teeth,

curved inwards, just like those which every one has seen in cats. These are the canines, which, as you have already learned, are always found in carnivorous animals.

3. They are teeth very well fitted for killing prey. If you could see the back teeth, too, you would see among them some

large teeth with sharp blades, well fitted to cut flesh; and the cat has teeth of the same kind.

4. You may see a lion in a wild-beast show without seeing its claws. The paws of the lion are soft and velvety like those of the cat; and the lion's claws, like the

cat's, are mostly drawn back and hidden in sheaths. But you know what sharp claws a cat can put out when it pleases.

5. The lion has claws of just the same kind, with sharp points and sharp edges, so that they can easily be buried in the soft flesh of any animal which it seizes. To each of these claws there is an elastic band, which holds back the claw when it is not in use; but both the lion and the cat have strong muscles, by means of which they can pull forward the claws when needed.

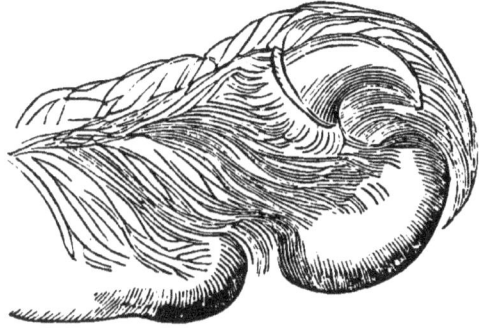

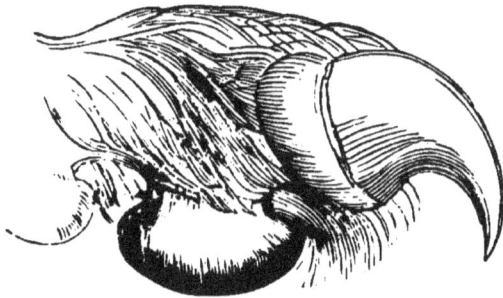

6. The eyes of the lion are also like those of the cat. Probably every boy and girl has

seen how much the black part in the middle of the eye, the pupil as it is called,
changes in size in the cat. When the light
is strong, you see only a narrow slit; but in
the evening, when there is little light, you
see the pupil wide open, and nearly round.

7. It is through that change that the
cat can see so well in the dark. It is that
which makes pussy able to catch mice in
the dark, when the mice leave their holes.
Now, it is just the same with the lion. The
lion, also, is an animal that seeks its prey
by night; and by means of its large pupils it
can see when it is almost quite dark to most
other animals.

8. The lion has long whiskers on the
sides of the lips, just as the cat has. You
would hardly guess of what use these whiskers are. But you know, that if you touch one
of these long hairs beside pussy's mouth, however gently, she at once feels it. She shakes
her head, and turns it away. Now, the whiskers often let pussy know, even when it is
quite dark, if there is anything near.

9. The lion, too, is warned in the same way; and when he is stealing along as silently as he can, he turns aside as soon as he gets such a warning, lest he should brush against anything, and make a noise which would startle the animal on which he means to make a meal.

LESSON XXIX.

rough	chim′-ney	fon′-dled	chief′-ly
mane	cov′-er-ed	sup-pose′	sup′-ple
tongue	knock′-ing	li′-on-ess	fe′-male
col′-or	slight′-est	cloth′-ed	des′-ert
taw′-ny	or′-na-ment	shoul′-ders	dwell′-er

THE LION.—2.

1. WE have not yet got to the last of the points in which the lion is like the cat. A girl who has fondled a cat upon her knee will often have had her hands licked by her pet, and she will know how rough pussy's tongue is.

2. The lion's tongue is rough too, and

indeed very much rougher than a cat's. The tongue of a lion is covered with hard, horny points, which are sharp enough to tear the skin on a man's hand when a lion licks it.

3. In the shape of the body, the cat and the lion are not very like each other. The lion has a very large head and high shoulders; and both the head and shoulders appear much larger than they are, because they are clothed with long hair, forming what is called a mane.

4. One can see more likeness between a cat and a female lion or lioness; for she has no mane, and has not so large a head nor such high shoulders as the lion. Yet even the lion has a very supple body like that of the cat.

5. The lion can twist and turn its body just as a cat can; and as pussy can walk along the mantle-piece without knocking over one of the ornaments, or making the slightest noise, so can the lion twist and

wind among trees and bushes, and yet make no noise by striking against them.

6. But you are not to suppose that the lion lives chiefly amongst trees and bushes. The lion is a dweller in the desert for the most part. As a rule, it has its den amidst rocks, in plains covered with sand or dry grass. On such ground its color helps to keep it hid: for the lion is always of a tawny color, not unlike the sand of the desert.

7. This tawny color is just of as much use to the lion. in helping it to come near its prey without being seen, as the white color is of use to the polar bear on the ice and snow.

LESSON XXX.

stalk	warn′–ing	sel′–dom	rap′–id
hurls	pur′–pose	ut′–ters	weigh
scent	light′–ning	hun′–gry	tim′–id
prey	ter′–ri–fied	ter′–ror	crunch
seize	care′–ful–ly	a–fraid′	chase
crawl	trav′–el–ling	stealth′–y	hoof′–ed

THE LION. — 3.

1. THE lion always draws near its prey as a cat does, in a very stealthy manner. Though a rapid runner, it does not run so quickly as the animals that mostly form its prey. These are almost all hoofed animals, which are very fleet of foot.

2. Sometimes the lion will lie long on one spot without moving, watching for some passing animal, as a cat watches at the hole of a mouse.

3. But very often it has to stalk its prey. That means that it tries, without being found out, to get so near an animal that it can dart upon it with a single great bound. In doing

so, it crawls along carefully on its belly, and shows great skill in hiding itself behind every bush, and every swelling in the ground, which can serve its purpose.

4. It always goes forward in such a way that the wind blows towards it from the animals it is watching ; for otherwise its scent might be blown towards them, and give them warning of its coming.

5. At last, when near enough, it makes its spring. It hurls itself in one great leap against the animal which it has singled out ; and unless the prey is very large, it is at once overthrown by the shock. For the lion is large enough to weigh nearly as much as three men ; and hardly any animal is able to bear up against the shock of such a mass, hurled against its side.

6. If the prey is not at once thrown down, the lion tears it down with his sharp claws, and kills it with a blow of his paw. The paw is strong enough to break the back of a horse with a single blow, and the lion's

jaws and teeth are strong enough to crunch the bones in the neck of a cow.

7. Often the lion will manage to secure a meal by uttering the awful roar for which

it is famous, and which strikes terror into the breast of nearly every beast of the plains.

8. Like all beasts of prey, the lion is much afraid of fire; and parties of men, when trav-

elling over the plains where lions abound, often try to save themselves and their cattle at night from lions, by making a ring of fire round their little camp. The lion dares not cross this ring of fire, but it utters its roar. Then the timid animals, struck with fear, and not knowing that they are safe where they are, rush out of the ring into the very danger which they dread.

LESSON XXXI.

straight	el'-e-phant	chil'-dren	ea'-si-ly
bis'-cuit	mam'-mal	an'-i-mal	height
al'-most	scarce'-ly	col'-umns	prop'-er
throat	drag'-ging	cu'-ri-ous	four'-teen
gi-raffe'	her-biv'-o-rous	im-mense'	pict'-ures

THE ELEPHANT.—1.

1. THE Elephant is the largest of all mammals that live on land. There are two kinds of elephants, one living in Africa, and the other in Asia. The first is the larger of the two, and sometimes reaches a

height of fourteen feet or even more, which means that it is more than twice the height of the very tallest men.

2. But it is not only great in height, it is big every way. Hardly any children need

to be told what the animal is like. Pictures of it they have often seen ; and if they live in a large town, they may have seen the animal itself walking the streets, and dragging a large car behind, when a circus enters the town.

3. In that way they have got to know its large, thick body, its straight legs like columns, its round feet with no proper toes, but with five small hoofs where the toes should be, its large, round head, its immense ears, its tusks, and above all its trunk.

4. This last part of the animal is well worth looking at. No other animal has anything like it. It is the elephant's hand and arm. It is that by which alone it gets its food and its drink.

5. The elephant is wholly herbivorous. It lives on the leaves of trees, or young twigs, and even on the bark and wood of small branches. Such food is mostly found high above its head. Grass and other herbs which grow on the ground are also eaten by it, but even that food it could not reach without its trunk.

6. For look where its head is. There is scarcely any neck, because a head so large could not be carried at the end of a long

neck. If you look at the picture facing the title-page, you will see near the elephant the figure of another animal that lives on the leaves growing on trees high above the ground.

7. That animal is the giraffe, and you will see in the plate how it is able to reach these leaves by means of its long neck and long tongue. But that animal has a very small head, easily carried. The elephant is different. It cannot raise or lower its mouth to find its food; but the trunk takes the place of the long neck of the giraffe, and has many uses besides.

8. The elephant can turn and twist its trunk in any way it pleases. It can wind it round young trees, and root them up. It can reach up to high branches, and tear them down. It can pluck up grasses and herbs from the ground, and lift them to its mouth.

9. When it wants to drink, it sucks water into its trunk, which is hollow, and then it pours this water down its throat. It can

even, by means of its trunk, pick up very small objects from the ground. For the end of the trunk can be opened and closed like a mouth.

10. The trunk has lips, so to speak; and those who have fed elephants with biscuits know how it makes use of these lips to hold small things given to it. But more than that, there is in the middle of the upper lip of the trunk something like a soft finger, which is always sticky, and by means of that finger the elephant can pick up almost anything that we could pick up between the finger and thumb.

LESSON XXXII.

for'–est	trop'–i–cal	al–though'	gen'–tle
thirst	bath'–ing	cli'–mate	at–tac k'
thongs	fu'–ri–ous	marsh'–y	in'–sects
roams	strug'–gle	con–sole'	nat'–ure
use'–ful	be–long'–ing	ser'–vant	bel'–low
en'–e–my	tor–ment'–ed	u'–su–al–ly	cap'–tive

THE ELEPHANT. — 2.

1. ALTHOUGH it can stand a climate as cold as our own, the elephant lives chiefly in tropical countries, or countries near the tropics. It roams about in herds in the forests, and is mostly found in marshy tracts, or at least in places where it can get plenty of water. For it is fond of bathing and washing its body with the water which it takes up into its trunk.

2. It even has the means of storing up water in its body; and when there is no other water near, it will put the end of its trunk in its mouth, and fill it with the water thus stored up, and then pour it over itself.

3. This it does to get rid of insects; and

in their own country elephants nearly always have certain birds on their backs which help to rid them of the many insects by which they are tormented.

4. Like most other herbivorous animals, the elephant is of a gentle nature, and never uses its strength to attack any other animal. On the other hand, it is too strong to be in danger of being itself set upon, even by the lion or the tiger. Its only enemy, in fact, is man; but the elephant is of too great use to man in many ways to be let alone by him.

5. In the first place, the elephant, being easily trained, can be made a very useful servant to man; and it is for this reason that the elephant belonging to Asia is usually caught. This elephant is mostly caught alive.

6. Men skilled in the work go out into the forest where elephants abound, taking with them tame elephants, which know very well what their masters intend, and know too how

to help them. The tame elephants go near the wild ones, and try to keep them from seeing what the men are doing.

7. The men thus manage to fasten ropes or thongs round their legs, and tie them to trees. The animals on finding themselves

caught become furious. They pull with all their might, they throw themselves on the ground, and twist and turn about in all ways, they bellow with rage, they tear up the ground with their tusks.

8. The struggle may last for several days.

But all in vain. The ropes and the trees hold firm. At the end of a few days, when the animal is worn out with its struggles, as well as by hunger and thirst, the men return with the tame elephants which console the captives, and afterwards help in training them to their duties. Such tame elephants are taught to do all that horses can do, both in peace and war.

9. The African elephant is never caught in this way; but it is killed in great numbers for the sake of its tusks, from which we get ivory.

LESSON XXXIII.

gi'–ant	whale'–bone	ter'–ri–ble	de–stroy'
pierce	wal'–low–ing	pur'–pose	nos'–tril
yield	rhi–noc'–e–ros	fright'–ful	pre–fers'
bul'–let	at–tack'–ed	med'–dles	i'–vo–ry
na'–tive	her–biv'–o–rous	trop'–i–cal	mon'–ster

THE RHINOCEROS AND HIPPOPOTAMUS.

1. THE two giants that you have to read about in this lesson, are in some things like the still larger giant that you read about in the last two lessons. All three live in the

warmer parts of the world, chiefly in tropical countries; they are all h e r b ivorous, and t h e y all have thick hides without hair.

2. You would hardly think that such ugly and frightful looking animals lived only on herbs, the leaves of trees, and other things got from plants. They seem made to destroy and kill.

3. Yet the horn of the Rhinoceros, and the tusks of the Hippopotamus, are chiefly used for rooting or plucking up plants, and neither the hippopotamus nor the rhi-

noceros ever meddles with other animals.

4. As a rule, both animals flee from man; but, as you might expect, they may both become very terrible when roused by being attacked.

5. The rhinoceros may grow to a length of thirteen feet, with a height of six feet. It mostly lives in the same kind of home as the elephant; that is to say, it likes marshy forests, and is very fond of wallowing in mire. One kind, however, lives on grassy plains.

6. The horn of the rhinoceros is very

unlike that of either the cow or the reindeer. When cut through, it looks somewhat like whalebone, and it grows only from the skin. When the animal dies, the horn falls off with the skin; and the skull, or bony part of the head, only shows a rough, thick piece of bone above which the horn stood.

7. Some kinds of rhinoceroses have two horns; but these horns are not placed at the sides of the head as in other animals, but one behind the other.

8. The hippopotamus is still larger than the rhinoceros. A large animal of the kind may be fifteen feet in length, and may weigh two and a half tons. It lives mostly in rivers; and to that it owes its name, which means "river horse."

9. It is a good swimmer and diver; and when it swims, it mostly keeps almost the whole of its body under water, showing only its ears, eyes, and nostrils. Like other mammals that live chiefly in the water, it can close its nostrils in diving, so that no water can enter them.

10. The hide of the hippopotamus is very thick. On the back, it is as much as an inch and a half in thickness. That of the rhinoceros is also thick, and is even tougher, so tough, indeed, that a common bullet will not pierce it.

11. A hard kind of bullet has to be made for the purpose of shooting this monster. The natives of the countries where the rhinoceros lives use its hide to make shields. The tusks of the hippopotamus yield a fine white ivory.

LESSON XXXIV.

fawn	stran'-ger	gal'-lops	stretch
twist	shoul'-ders	slen'-der	pret'-ty
tongue	beau'-ti-ful	cu'-ri-ous	gi-raffe'
clumps	any'-where	cov'-er-ed	quick'-ly

THE GIRAFFE.

1. Is there a stranger-looking animal to be seen anywhere? One might think not, on first seeing a Giraffe; yet so many and so curious are the forms of animals, that it would not be right to say that even the giraffe is the strangest of all.

2. No one who sees a giraffe can help being struck with its long neck, the small head at the end of it, the long, slender legs,

and the very high shoulders causing the back to slope so much downwards to the root of the tail.

3. It cannot be called a beautiful animal. It has, indeed, a beautiful coat of a light fawn color, covered everywhere above the knees with brown spots. The eyes, too, are

large and fine; but the shape of the animal spoils its beauty.

4. Is there any use for this very long neck? Yes; there is. It is by means of that long neck that the animal is able to reach the food on which it likes best to feed. The giraffe is an animal that chews the cud, like the ox or bison; and it lives on plant food as all such animals do. But its food is not grass. It is the leaves of trees, leaves that grow high above the ground.

5. Giraffes live in small troops in very dry parts of Africa, where there is very little grass, but where there are clumps of trees here and there. By means of its long neck, the giraffe can reach leaves twenty feet above the ground; and these it nips off with its long tongue, which it can twist round the small twigs.

6. It hardly ever feeds on grass, and indeed cannot easily pick up anything from the ground. In order to do so, it has to stretch out its forelegs as wide apart as it can, and then to lower its neck.

7. You will see that the giraffe has small horns, but these are not like those of the cow or bison. They are not covered with horn, but are made of bone covered with skin and hair, like the rest of the body.

8. The giraffe can run pretty quickly; but it never trots, it always gallops. In doing so, its long neck, which it nearly always holds stiff and upright, keeps swaying from side to side like a long pole.

LESSON XXXV.

loose	creat'-ure	up'-wards	ac-count'
smooth	ug'-li-ness	jut'-ting	en-a'-bles
cam'-el	dis'-tan-ces	car'-a-van	parch'-ed
shag'-gy	yel'-low-ish	marsh'-es	cush'-ion
bur'-den	un-grace'-ful	for'-wards	swol'-len

THE CAMEL. — 1.

1. No one who has seen a Camel could think that the giraffe is the strangest-looking of all animals. What living creature is more unshapely than this? Everything about it seems ugly.

2. Look at its broad, swollen feet, its
thick joints; its long neck, first jutting forwards, and then rising upwards in a very

ungraceful manner; the ill-shaped head, with
the long loose lips; and above all look at the
great hump on its back.

3. Even its color has little beauty in it. It is a plain yellowish brown; and the hair is in some places short and pretty smooth, but in other places it forms shaggy tufts.

4. But in spite of all this ugliness, no animal is more highly prized by the people among whom it is found. There is no animal better fitted for the life it leads; and on that account the people of the countries to which it belongs, would find it as hard to live without the camel as the Lapps without the reindeer.

5. How different are the homes of these two animals! In the one case, a country of snow in winter and marshes in summer; in the other case, a country in which there is only burning heat all the year round, in which rain never, or hardly ever, falls, and in which the soil is a parched sand.

6. In that sandy country, the camel is almost the only beast of burden. No other animal can be used to carry loads for long distances across the desert. The desert is

crossed by large companies of men with laden camels, forming what are known as caravans. When people are divided from each other by broad deserts, it is by means of such caravans that they carry on their trade. Hence the camel is often called " the ship of the desert."

7. Almost everything that has been pointed out as strange and ugly in the animal helps in fitting it to serve this use; and there are other things, still to be mentioned, that do the same.

8. You were asked to look at its feet. Look at them again, and let us see how these help it to lead its desert life. If you look carefully, you will see that each foot has two hoofs, and you know therefore that the camel chews the cud, and is an herbivorous animal.

9. But how different this foot is from that of the cow or giraffe. The hoofs in the camel make up only a small part of the foot; the greater part is a soft pad or cush-

ion, with a tough skin underneath. That enables the camel to walk without pain for a long time over the sand of the desert, and to stand its burning heat.

———◆——

LESSON XXXVI.

fare	car´-a-van	swol´-len	scarce
wart	pres´-ence	con-tent´	cam´-el
scent	al-to-geth´-er	en-a´-ble	des´-ert
quart	her-biv´-o-rous	stom´-ach	flab´-by

THE CAMEL.—2.

1. As the camel is herbivorous, you may ask where can it get food in the desert. Well, it is content with very poor fare. Large tracts in the desert are without any plants at all; and where plants do grow, they are very often prickly shrubs, with thorns strong enough to pierce the sole of a man's boot when he treads upon them.

2. Yet upon these shrubs the camel can make a good meal, and the thorns seem to do no harm to its mouth or tongue. Grass,

where it does grow, is often very short, yet
the camel can nibble it with its long, loose
lips.

3. Where food is very scarce, the camel
can go longer than any other four-footed
beast without
any food at all,
and what ena-
bles it to do so
is that great
hump on its
back.

4. That hump
is nearly all fat.
When the camel
is well fed, the
hump is large
and firm ; but when it has been much in
want of food, the hump becomes loose and
flabby, and begins to hang down. Sometimes it almost goes away altogether, because the fat in it takes the place of food,
and is used up in keeping the animal alive.

5. The camel must drink too, and water is as scarce in the desert as food is. But the camel can go as long without water as without food, and the reason for this is a very strange one. In the walls of that part of the stomach of the camel into which the food is first taken, there are a great many little bags or cells all in rows.

6. There may be eight hundred or more of these; and each of them can be closed at the mouth, very much as a bag is closed by means of a string passed through a fold round the mouth and drawn tight. Now, in these cells the camel can store up water for use, in times when it cannot get water to drink.

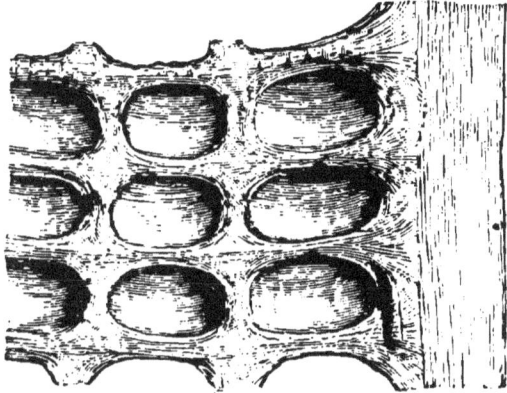

7. It can store up five or six quarts at

once, and then it can live for more than a week without a fresh supply. The camel can tell the presence of water a great way off by the scent, and it takes in a fresh store whenever it can.

8. When resting or being loaded, the camel lays itself down on its belly with the legs turned in under the body. So that it may not be hurt in so doing, its breast and the knees of its forelegs have hard patches or warts over them; and it is the warts of the knees that give such a swollen and ugly look to the joints.

9. A camel load weighs over 500 pounds, so that it would take 5,000 camels to carry as much as a single large ship. From that you can see how much better for trade the ship of the sea is than the "ship of the desert." A large caravan has, in fact, thousands of camels.

LESSON XXXVII.

breed	un-ea′-sy	slen′-der	rap′-id
cream	hand′-ful	jolt′-ing	draught
stretch	care′-ful-ly	car′-ri-er	rear′-ed
yoked	nour′-ish-ing	jour′-ney	wov′-en

THE CAMEL. — 3.

1. SOME breeds of camels are carefully reared to be used only for riding on, as with us some breeds of horses are reared for racing or hunting. Such camels are of a finer kind than the common camel. Their legs are longer and more slender, and so their speed is much greater.

2. A riding camel will go at the rate of eight or ten miles an hour, and on a long journey is a much more rapid animal than the horse, for it can keep up this rate for twenty hours at a stretch. The camel can thus make a journey of nearly two hundred miles in a day.

3. It is one of these camels, mounted by a letter-carrier, that is shown in the cut on

page 128. and there you can see how the camel is ridden.

4. The gait of the animal causes so much jolting that the rider's seat is a very uneasy one. It has been said, that one reason for calling the camel the ship of the desert is that people are sometimes "sea-sick" when they first ride on one.

5. To the people of desert countries, the camel is of use in other ways. Its flesh is seldom eaten indeed, because the animal is of too much value to be killed for food. But the milk of the camel is very nourishing, and when mixed with meal is a dish which the people of the desert like very much.

6. A kind of butter is made by shaking the cream in a goatskin bag. To us the taste of this butter would not seem at all

good, but the people who make it are very fond of it.

7. The long hair of the camel is spun and woven into cloth. At certain times of the year, the camel sheds its hair as many animals do; and then the long hair can easily be plucked off in handfuls, without doing any harm to the camel.

8. The camel that is shown in our wood-cut has only a single hump, and it is that kind of camel that is mostly used as a beast of burden.

9. But there is another camel which has two humps. It belongs to the desert parts in the middle of Asia, and is more used as a beast of draught than as a beast of burden. That means that it is yoked to carts instead of being used to carry loads.

LESSON XXXVIII.

prey	trop´–i–cal	ser´–pent	throat
fangs	swal´–low	cu´–ri–ous	stretch
limbs	di–gest´ed	wrigg´–ling	e´–qual
height	skel´–e–ton	squeez´–ed	mo´–tion

THE BOA CONSTRICTOR.

1. THE Boa Constrictor is one of the largest of serpents or creeping animals. Such animals have no limbs whatever. They have long, round bodies, and their skeleton behind the head is only a long back-bone with ribs joined to it.

2. They move on with a wriggling motion; yet they can do it with such force, that it is not easy to pull even a small snake backwards. You will wonder how this can be. And the reason is very curious.

3. Instead of having limbs to move, the serpent can move all its ribs forwards and backwards. In moving them forwards, a number of plates or scales that stretch from side to side on the under part of the body,

also move in such a way that they are all
set on edge. The edges catch hold of the
ground, and then the serpent can pull its
body forwards until the scales lie flat again.
Often, too, they draw themselves up by the
same means into great folds or loops, and
then they move all the more quickly.

4. The serpent that you are to read about
in this lesson lives in tropical countries. Its
home is the tropical part of America. It
grows to a length of twenty feet or more,
which is equal to three or four times the
height of a tall man. Its color is a rich
brown. with rows of black and white spots
on the back. Like all serpents, it has its
body covered with scales of a smaller size
than those which are under the body and
joined to the ribs.

5. Though they have a backbone, serpents can twist and wind their long bodies
almost as freely as worms do, which have
no bones at all. Some of them make use of
this power to kill their prey, and the boa is
one of these.

6. It coils itself two or three times round an animal which it means to eat, and having done so, draws the coils tighter and tighter with very great force, until the poor thing is not only killed, but squeezed into a form in which the boa can swallow it whole. For all serpents have that strange way of taking their food. Their teeth are of no use for dividing up what they eat. They are merely strong, sharp pegs all pointing backwards, so that an animal once caught

by them has no chance of slipping away again.

7. The serpent takes hold of its prey by an end thin enough to be taken into its mouth. and slowly forces it down its throat. And the bodies swallowed are often much thicker than the serpent itself. For the bones of the skull in the serpent are all loose, so that the mouth as well as the body of the serpent can stretch a great deal. The boa, though only a few inches in thickness, can swallow animals as large as a sheep or a goat.

8. After a great meal like this, the boa lies as if in a deep sleep for weeks; and not till its heavy meal has been fully digested, does it wake up again. While in this state it is of course very easily killed.

9. Some serpents kill their prey by means of poison, which they squeeze into wounds made by long teeth called fangs; but the boa is not one of these.

LESSON XXXIX.

shelf	car'-ri-on	ac-count'	pre-fer'
range	wheel'-ing	soar'-ing	drear'-y
swoop	stretch'-ed	meas'-ure	al'-most
height	ev'-er-y-where	grace'-ful	cer'-tain

THE CONDOR.

1. THE Condor is the largest of flying birds. It is more than four feet in length from the tip of its beak to the end of its tail; and its wings, when stretched out to their full length, measure about nine feet, sometimes a good deal more than that.

2. Like the other animals you have lately been reading about. it lives in tropical countries; yet its home is very much unlike what you would think from the account of these countries which you read in a former lesson. The condor is a bird of the highest mountains; and you will remember being told, that even in tropical countries there is snow at a great height.

3. Now, if you look at South America

on the globe, you will see in the west a
line or range of mountains running the
whole way down from north to south.

These mountains are almost everywhere so
high that the snow never melts all away
from their tops, and it is there that the
condor has its home.

4. Among these silent and dreary heights, where hardly any other sign of life is to be seen, where the thundering of falling masses of snow from time to time is almost the only sound that is to be heard, the condor lays its eggs on a bare shelf or rock, without building any nest.

5. Its food is the flesh of animals. It is a carnivorous bird, like the eagle and the hawk. Like them it has a strong hooked beak by means of which it can tear the flesh of its prey. Mostly it prefers the flesh of animals that have already been killed ; but sometimes it will attack a living animal, first blinding it by tearing out its eyes, and then killing it with blows of its beak.

6. The food of the condor cannot of course be got near to its home. But the bird has a keen eye, and from its lofty standpoint can make out a dead animal a very great way off. It then swoops swiftly down, and feeds itself on the carrion.

7. Often the condor can be seen very high

in the air, nearly six miles as some say, soaring over a certain spot in graceful curves. It may then be watching a dying animal, or a four-footed beast of prey attacking a weaker animal, for a share of which it waits.

8. Sometimes the condors are in small flocks, and the sight of such a flock wheeling round and round any spot has been found to be well worth watching. One who watched them closely states that he does not remember ever to have seen these birds flap their wings, except when rising from the ground. They seem to glide up and down merely by moving the head and neck. For hours they may be seen wheeling and gliding in this graceful manner, over mountain and river.

9. When on the ground, it is often easy to get near a condor and to kill it; for it cannot rise into the air unless it has room to run some length along the ground; and the men who are after them always try to give them no chance of doing so.

LESSON XL.

hawk	en′-e-mies	ex-pos′-ed	par′-rot
breathe	re-quir′-ed	trop′-i-cal	val′-u-ed
dain′-ty	hawks′-bill	u′-su-al-ly	hatch′-ed
hook′-ed	leath′-er-y	At-lan′-tic	tor′-toise
sur′-face	pro-tect′-ed	rel′-ish-ed	Pa-cif′-ic
pad′-dles	am-phib′-i-ous	fre′-quent	cu′-ri-ous

TURTLES.

1. TURTLES are amphibious animals like beavers, living partly on land, partly in the water. Beavers, however, have their home on dry ground, while turtles are nearly always in the water. Some animals like turtles frequent rivers, as beavers do; but true turtles live like seals, another kind of amphibious salt-water animals, in the sea.

2. They are mostly found in tropical seas, not very far from land; though sometimes they are to be seen hundreds of miles from any shore. The land which they keep near is usually a small island, for on the mainland the young are exposed to too many enemies.

3. The shape of a turtle is very curious.

The body is very broad and rather flat, and
most of it is protected by plates of bone, both

above and below, those above being in the
form of an arch.   Above the bony plates on

the back, there is also a horny or leathery covering; and beyond this the head, limbs, and tail jut out. The limbs are shaped like paddles, and by means of them the turtle can swim well.

4. Though living mostly in the sea, they cannot breathe under water as fish do. They must come to the surface to breathe, like whales and seals. Yet they are not mammals. They do not suckle their young with milk. They lay eggs not unlike those of hens; and the young, when hatched, feed themselves.

5. There are many sorts of turtles. They are all large. Some are as much as eight feet in length, and weigh sixteen hundred pounds. When found at all, they are found in great numbers. Near some small islands in the Atlantic and Pacific Oceans, the seas swarm with them.

6. Of all the many kinds of turtles. there are two which are much more highly prized by man than any others, but not for the

same reason. One of them is the green turtle; and it is valued for its flesh, which is looked upon as a great dainty. It is called the green turtle on account of the color of its

fat, and this green fat is the part that is most relished by those who are fond of turtle. In this country turtle flesh is so dear that it is eaten only by the rich.

7. One common mode of catching the turtle is curious. Men watch for them to come ashore ; and then, running between the turtle and the sea, they turn as many as they can over on their backs. Thus they lie helpless until the men return to carry them away. So large is the green turtle that three men are often required to turn one on its back.

8. The other kind of turtle which is highly prized is not eaten, for its flesh is not good; but its back is covered with large horny scales, from which the beautiful " tortoise-shell " is obtained. This turtle is called the hawksbill turtle, because its beak is hooked downwards like the beak of a hawk or that of a parrot.

PHILIPS'
GEOGRAPHICAL CHART

for Elementary 'Classes. One large sheet. Size, 68 x 54 inches. Mounted on calico, rollers, and varnished $5.00

Comprises the following : —

(*a*) A large Map of Great Britain and Part of the Continent of Europe. Illustrating the various Geographical Definitions, Political and Physical.

(*b*) A large Pictorial Scene, illustrating to the eye the chief Features of Land and Water.

(*c*) Diagrams of Schoolroom, Schoolhouse, and Ground Plan of School Buildings.

(*d*) Mariner's Compass.

(*e*) Pictorial View of the Course of a River, from its Source to the Mouth.

(*f*) Diagram illustrating method of ascertaining direction from the Sun — North, South. East, or West.

(*g*) Map of the Globe. showing Division of Land and Water.

(*h*) Six Typical Heads, illustrating the Races of Mankind.

(*i*) The Earth in Space.

(*j*) Diagram showing the Curvature of the Earth.

The above Chart has been prepared with great care, and will be found extremely helpful in class teaching.

PHILIPS'
GEOGRAPHICAL READERS.

THESE READERS have been most carefully prepared, and the publishers feel confident that in the treatment of the subject, the style and quality of the matter, the number and beauty of the illustrations, the legibility and accuracy of the maps and diagrams, the books will be found superior to any other similar series, and will render the study of geography interesting and attractive. The series contains no less than 800 valuable illustrations and maps.

1. **FIRST STEPS.** Part I., explaining "plans of school and playground, the cardinal points, and meaning and use of a map." With word-lists and summaries, 32c.

2. **FIRST STEPS.** Part II. "The size and shape of the world, geographical terms simply explained and illustrated by reference to the map of England, and physical geography of hills and rivers " . . . . 36c.

3. **ENGLAND,** Physical and Political, in a graphic narrative form . . . . 43c.

4. **BRITISH ISLES, BRITISH NORTH AMERICA,** and **AUSTRALASIA,** described in a series of well written sketches of voyages, travels, etc. . . . . . 65c.

5. **EUROPE,** Physical and Political, described in a series of narratives of voyage and tours. With Appendix — Latitude and longitude; day and night; the seasons . . . . 75c.

6. **THE WORLD.** A series of voyages and travels in Asia, Africa, America, and Polynesia. With Appendix — Interchange of productions; circumstances which determine climate 86c

THE INFORMATION READERS.

THIS SERIES is significant of the profound change which school methods and theories have undergone within the past decade.

The Information Readers are issued in response to the increasing demand for reading-books that, while enlarging the vocabulary of the young learner, shall tell him something of the busy work-a-day world around him.

1. **FOODS AND BEVERAGES,** by E. A. BEAL, M.D. Contains reading lessons on the various kinds of Foods and their hygienic values; on Grains, Fruits, and useful Plants, with elementary botanical instruction relating thereto; on the effects of Stimulants; and on other common subjects of interest and importance to all, old and young. 281 pages. Cloth . . . . .60c.

2. **EVERY-DAY OCCUPATIONS,** by H. WARREN CLIFFORD, S.D. Quantities of useful facts entertainingly told, relating to work and workers. How Leather is Tanned; How Silk is Made; The Mysteries of Glass Making, of Cotton Manufacture, of Cloth Making, of Ship and House Building; the Secrets of the Dyers' Art and the Potters' Skill — all and more are described and explained in detail with wonderful clearness. 330 pages. Cloth . . . . .60c.

3. **MAN AND MATERIALS,** by WM. G. PARKER, M.E. Shows how man has raised himself from savagery to civilization by utilizing the raw material of the earth. Brings for the first time the wonderful natural resources of the United States to the notice of American children. The progress of the Metal-Working arts simply described and very attractively illustrated. 323 pages. Cloth . . . . .60c.

4. **MODERN INDUSTRIES AND COMMERCE,** by ROBERT LEWIS, PH.D. Treats of commerce and the different means of conveyance used in different eras. Highways, Canals, Tunnels, Railroads, and the Steam Engine are discussed in an entertaining way. Other lesson subjects are Paper Manufacture, Newspapers, Electric Light, Atlantic Cable, the Telephone, and the principal newer commercial applications of Electricity, etc. 329 pages. Cloth . . . . .60c.

PHILIPS' HISTORICAL READERS.

1. **STORIES FROM ENGLISH HISTORY.** 128 pages; 38 Short Lessons, with numerous Explanatory Notes; 62 beautiful Pictures, and a Map of England and Wales. Price, 36 cents.

These stories from English history form one of the brightest and most attractive Reading Books ever published. Each story is not only well written, but also beautifully illustrated. The portrait of Her Majesty the Queen, which forms the frontispiece, is extremely fine. Altogether, this book is an admirable introduction to the study of English history.

2. **EARLY ENGLAND, from Pre-historic Times to the Year 1154.** 192 pages; 54 interesting Lessons with useful Notes; 94 attractive pictures; 6 finely engraved maps. Price, 42 cents.

In this beautifully illustrated and well-written little book, the story of the making and founding of the nation is graphically sketched. The opening section contains vivid pen and pencil pictures (based on the latest antiquarian and geological research) of life in that country in pre-historic times — the periods of the men of the caves, the stone-hatchet men, the bronze-workers, etc.

3. **MIDDLE ENGLAND, from 1154 to 1603.** 256 pages; Price, 62 cents.

In this book, the history of the country is continued from the reign of Henry II., when the welding of Saxons and Normans into one compact people commenced, to the end of the reign of Elizabeth, when the modern, social, political, and scientific ideas had at last been fully thought out. The aim has not been to give merely the "lives" of the kings and queens, or the records of war and victory, but to present, clearly and accurately, the real history of our English forefathers during what may be justly termed the *decisive period* of English history.

4. **MODERN ENGLAND, from 1603 to 1883.** 272 pages. Price, 62 cents.

In this book, the great events of the last 280 years are graphically and succinctly described and fully illustrated. The high educative value of good pictures has been constantly kept in view, and the number and beauty of the illustrations form one of the characteristic features of the Series. The greatest possible care has been taken to preserve an absolutely impartial tone throughout the Series.

NATURAL HISTORY READERS.

By the Rev. J. G. WOOD, M.A.
Author of " Homes Without Hands," etc.

THIS SERIES OF READERS is carefully graduated, both as to matter and language; the list of words for spelling is selected with due regard to actual experience of children's difficulties, and is therefore in every way fitted to serve the purpose of ordinary reading-books.

Nothing more readily interests children than animal life. It will be noticed that in the lower readers no animals are introduced but those that are more or less familiar to children; the subjects are treated in such a manner as to lead the way naturally to the scientific classification introduced in the higher books.

FIRST READER.
Short and simple stories about Common Domestic Animals . . . . 25c.

SECOND READER.
Short and simple stories about Animals of the Fields, Birds, etc. . . . . 36c.

THIRD READER.
Descriptive of familiar Animals and some of their wild relations . . . . 50c.

FOURTH READER.
The Monkey Tribe, the Bat Tribe, the Mole, Ox. Horse, Elephant, etc. . . . . 65c.

FIFTH READER
Birds, Reptiles, Fishes, etc. . . . . 65c.

SIXTH READER.
Molluscs, Crustacea, Spiders, Insects, Corals, Jelly Fish, Sponges, etc. . . . . 65c.

www.ingramcontent.com/pod-product-compliance
Lightning Source LLC
Chambersburg PA
CBHW021814190326
41518CB00007B/585